ENVIRONMENTAL PROBLEMS IN 21st CENTURY

ENVIRONMENTAL PROBLEMS
IN 21st CENTURY

By
I. Mohan

ANMOL PUBLICATIONS PVT. LTD.
NEW DELHI - 110 002 (INDIA)

ANMOL PUBLICATIONS PVT. LTD.
4374/4B, Ansari Road, Daryaganj
New Delhi - 110 002
Ph.: 3261597, 3278000
Visit us: www.anmolbooks.com

Environmental Problems in 21st Century

First Edition, 2002

ISBN 81-261-1312-X

PRINTED IN INDIA

Published by J.L. Kumar for Anmol Publications Pvt. Ltd., New Delhi - 110 002 and Printed at Mehra Offset Press, Delhi.

Dedicated to

My late father Shri Girja Dutt, my mother Smt. Dikeshwari Devi, my wife Mrs. Pushpa Mohan and Mrs. Shanoo and Mr. Rajiv Bhatt, Mrs. Deep Shikha and Mr. Pankaj Khanduri, Ms. Tanu Mohan, Masters Abhas Mohan, Mr. P.K. Hota, Shubham Khanduri and Shorja Bhatt.

as well as to

U.N. H.Q., World Bank, W.H.O. UNICEF and TERI

Contents

Preface ix

Acknowledgements x

List of Sketches xi

1. Introduction 1
2. Turbulent Himalayas 5
3. Shri Sunder Lal Bahuguna's Book Inauguration 10
4. Silent Fast Day of Shri Sunder Lal Bahuguna 12
5. Meera Bahan ka Prakriti Pyar 15
6. Creating Environmental Awareness 18
7. The Moghul Palace—Red Fort, Shahjahanabad 26
8. Forests 30
9. Aravali Ranges 34
10. River Banks: Yamuna River 52
11. Wild Life 67
12. Environment and Habitat 71
13. Implementation of Total Quality Management 76
14. The Poverty Alleviation Programmes and Environment 79
15. The Future of Waste Management Practices for Energy Recovery 86

16. Energy Conservation in a Developing Country 89

17. The Management Evaluation of Marine Pollution 93

18. The Protection of Wild Life 95

19. Environmental Upgradation Through Efficient Management of Water Resources in the National Capital Region 106

20. Energy Conservation in a Developing Country 124

21. Handling a Natural Disaster in a Developing Country 132

22. The Indian Tourism Industry 146

23. University Tours as Tourism Industry 151

24. The Project Management 155

25. Rural Structure in Uttaranchal Region: Need for Sustainable Development 170

26. Controlling Sprawling Rural Urban Villages in Delhi 176

27. Can Ever Our Rural Areas Improve? 179

28. Improving Environment of Rural Areas by Providing Necessary Facilities 181

29. Improving Urban Environment through Architects 186

30. A New Syllabus for Environment 191

31. Linking Environment to New Technology 193

32. Shahjahanabad is Everywhere 195

33. Mohan's Concept 202

Bibliography 219

Sketches 223

Preface

The author since last 30 years engaged to the subject, Environment, has been proposing diversified directions and is ultimately waiting to see a change. There have been strict laws framed up. Also demerits of various solutions are found in the past. So a confirmed framework, he feels, is required with a proper syllabus so that it may no longer become an unending show and thereby remain uncontrolled.

The technologies are required to take the place for each issue since computerisation operations are there. Therefore, much shall depend on linking Global neighbouring countries. A harmonised plan for the world needs to be chalked out.

The simplicity, the straightforward attitude and the transparency are the three main issues that will take up the place of science of environment. It is good that trees, plants, animals and biotic issues within the environment are linked to urban development for which the author has worked very hard.

The author desires to experience a Global tour and welcomes any country's request for a visit as advisor on this issue. For any sort of query he would be pleased to reply and could be consulted at his residential address.

10B, A.N. Shalimar Bagh,
Delhi-110088 India

I. Mohan

Acknowledgements

The author is thankful to all persons who have helped him in writing the book and offered him an idea of linking the environmental issues; the first of its kind with the total Globe in the hope that from the beginning all environmental issues shall be remonitored by each and every country suiting it.

He owes deep regards to the following persons, Mr. Sunil Dutt a film artist, Mr. Vijay Kapoor, Mr. Mr. S.K. Sharma former MD, HUDCO, Mr. Anil Kumar IAS, Mr. K.P. Notiyal former V.C., H.N. Bahuguna, University Sri Nagar, Mr. Sudripto Roy IAS, Mr. K.J. Alphons IAS, Mr. U.S. Jolly, Mr. V.M. Bansal (DDA). Also thanks are for Mrs. & Mr. Harish Bhatt, Mrs. & Mr. D.P. Khanduri, Mrs. & Mr. S.M. Malasi, Mrs. Kusum, Mr. Mukesh Malasi, Mrs. Suman and Mr. Rakesh Malasi, Mr. J.P.J. Bilmoria, Mrs. & Mr. Chander Mohan Gaur, Mrs. & Dr. Man Mohan Gaur, Mrs. Kanta and Mr. A.K. Uniyal and Mrs. Usha & Mr. D.P. Dangwal.

The author exclusively remembers late Mr. K.D. Mistry, Professor, School of Planning and Architecture, Delhi, whose inspirations are still in the mind of the author for better expression in drawings such that those may speak out.

—I. Mohan

List of Sketches

1. Plan Showing Aravali Ridge and Yamuna River at Delhi 223
2. Sketch of Slum Area 224
3. The Old Schools/Colleges/University for Tourism Development 225
4. Linking Future Cities with Villages 226
5. Typical Rural Housing Plan 227

1

Introduction: Next to Environment What?

Ever since the Global awareness about environmental protection was raised by the affluent nations, the developing countries took that in a serious way with the result of which earlier this remained confined to a few issues like afforestation and protection of wild life. Gradually that took a large dimension when urban development issues got clubbed to that alongwith other Global sciences aiming at protection of the human beings from impending ecological disasters like the social constraints pertaining to various human settlements, unemployment, nourishments and exceeding stress on pollution of air, water and wastes.

As talks were more conservation oriented, at the same time the factors responsible for polluting environment got multiplied which exerted influence on authorities to sort that out and that affected many in terms of livelihood more so who lived in slums. The three categories of people reacting with each other in slums—are the bureaucrats, the politicians and the public for which it is predicted that the system can never be improved even for centuries.

The Apex Court's intervention in the system is a one side story which is pressing hard the municipality's bureaucrats to sort out various issues and to a large extent is good also but at the same time due to lack of manpower, police power and other constraints, this has no doubt created a fear among bureaucrats who have to attend such jobs on priority leaving the routine work kept pending on their tables.

It's a good sign as is observed that the Ozone layer depletion in nature is healing up as a result of which the countries which stand along sea will get relief. The coastal pollution due to waste collection on beaches is also improving as International laws are framed up banning dumping of the garbage directly into the sea. There is a large ecological balance in the sea about which there is a great concern of saving the lives of sea animals from pollution.

The pollution of air, water, noise and wastes have been taken care of to protect the public from the ill effects with the result that there is a resentment as this is disturbing the social life, as such, measures are expensive and sometime lead to demolition or closing of job units creating thereby unemployment and affecting the people. The industries which were earlier located on the periphery of the city have now come up within at the centre and if that is there then it is difficult to get rid of that. In Delhi, The proposal in Master Plan was to place those industries outside the city and ceiling of such industries within the city. Uprooting an established business creates problems socially, but from the point of view of health controls may result breaking out of diseases and epidemics. Next to Industries are the traffic issues creating air pollution. No doubt the vehicles are multiplying and there is no control and large showrooms of multinational companies are daily coming up with priority of CNG system but how far this would be successful cannot be gauged but definitely this would meet resistance.

Let's go deeper in the situation across conservation of heritage, trees and wild life. It is felt that those have wrongly been backed up. The laws are also so strict that no one desires to look at the truth. We see that the panthers create havoc in rural and urban areas, a number of dogs and monkeys bite civilians in urban centres and growth of trees on sun side block sun to the residents in the winter.

The fight between poachers and authorities are also creating problems in administration. At one point of time number of tigers at site multiplied so much that no infrastructure was available for accommodating them with the result that they all fell prey to epidemic. The laws are so tailored that no body can help.

The above version in fact raise a question of ending one system and starting another a better one, reliable and more justifiable. What that moment could be said but would require a definite issue connected to advance technologies? Would that be an urban issue, biotic issue or project issue shall have to be seen.

The Global strategies are not synonymous and vary in population, climate, economy and social systems. A definite living style every one has so far formed. Then to have with them Global similarity as well as for individual, more stress is required for the later. What should be the criteria then; very simple, the individual human being, his health should not be effected and he should enjoy all his life. Man, Man and Man should be the motto says Mohan's concept which need be covered. If there is natural disaster let the residences be built of wood or bamboos, no harm. Atleast life will be safer and free of risks.

Let me know despite alarming signs of disasters has there been any change in the system? No. Why so, cannot be said but now it has to be expressed as the helplessness of public who watched the scenes. That can be said to be related to the environment and human beings around the world. Such attitude like a project need be defined. Even anticipation of that after 10 years be also told, magnified, evaluated and projected for 100 years.

Let's close or review next to environment, what I think environment is appearing to be monotonous and requires replacement by technological inventions. The changes should take care of all needful environmental issues. There is nothing impossible.

There is required to be homogeneity of the developments so that the difficulties can be well guessed of a country if at all there is. A ring of countries need linkages to develop as good neighbours. The planned development shall be on a same size of square (land) alignment repeated all over. The major roads of a country need linkages to other country's roads. This has perhaps never been done. The countries so joined could be called Mohan's International Regions.

The project may formulate for maximum 10 years and be very accurate and for rest can be just given in outline subject to changes of constraints; population and environment. But if that remains confined to biotic issues that would be better.

The new technologies would play a major role and multinational companies need to participate in that. Every system would require enhancement and be humanely oriented. Atleast the life would be like doing away with diseases, epidemics that take some times heavy toll of lives. Still human life has vastly improved due to the advancement in medical science. These can provide employment and shelter to the people. The European and affluent countries have made life luxurious and they do not have to depend on other agencies for help. The self styled machines do the work on orders as is given to ladies at home. The life will become less burdonsome. This field will become so improved, that rich companies would start working on the micro issues and engage public to do more research in that.

For example on a Neem tree, in India, research can be done on bark, leaf, fruits and bonsai and let there be special institutes for each one of those, says Mohan's concept. That would be a scenario of forthcoming 50 years.

2

Turbulent Himalayas

The Himalayan region always makes headlines during the monsoon, when huge landslides occur and highways are blocked. This is now regarded as a frequent phenomenon. But more than road blockades, on July 25, massive rocks rolling down, not only blocked the Rishikesh-Gangotri highway but the Bhagirathi itself was blocked at Loharinag which is located 52 km upstream from Uttarkashi. The event would have remained unreported, had not about 200 pilgrims returning from Gangotri, been marooned at this spot.

An artifical lake was formed due to the rockfall, which posed a danger to habitation downstream. Another landslide at Dabrani nearby took 11 lives, including those of eight army personnel.

The blockades of Bhagirathi at Loharinag and Dabrani are not isolated events. During the last 33 years, such blockades have occurred repeatedly at Loharinag itself. In 1959, in a major disaster the road itself was washed away. The lone survivor was a horse, which could be seen grazing by the lake side for many days. It was with great difficulty that road communication could be restored.

In September 1978, in another major disaster, a lake was formed at Dabrani which threatened the entire Ganga valley. The people were cautioned to be alert right up to Hardwar.

Courtesy: Echo from the hills by Sh. Sunderlal Bahuguna

When this artificial lake burst, it washed away habitations and fields on both sides of the river in Uttarkashi and Tehri districts. Road communication could be restored only after two years.

The occurrence of landslides in the central Himalayan region during the past two decades have taken place with unfailing regularity. The Belakuchi mishap in Alaknanda in July 1970. Tawaghat landslide in 1977 and thereafter several disastrous landslides in Almora, Chamoli, Tehri-Garhwal and Uttarkashi districts, destroyed many villages and buried many human beings and livestock alive.

There is even a longer list of such mishaps in the region which have not been documented. All these combined have, in fact, completely changed the landscape of the area. Uttarkashi at one time was on the left bank of the river, but due to floods, the river changed its course and now it is on the right bank of Bhagirathi. Similar changes occurred in the landscape of the Gangotri Gomukh region.

An old Garhwali folk song gives a vivid description of a huge landslide near Jhala. Jhala is 15 km upstream of Loharinag. The hillock of Dhaini near Jhala fell down and blocked the river. It formed a 10 km lake from Jhala to Jangia. Till recently nobody believed this had happened. The truth was revealed when an old Shiva temple was excavated in Dharali village. The folk song mentions the existence of 240 temples in Dharali, the last site of habitation on the way to Gangotri.

The Uttar Pradesh Government at one time appointed a committee to suggest ways and means to check landslides, but its report was buried forever as it contained critical remarks about the development policy and suggested some radical changes. Similarly, independent scientific reports on the Kanodiagod landslide (1978) were ignored. Planners and decision-makers in the field of water resources do not even like the mention of these reports for fear of their ill-conceived projects being questioned.

Prof. Kharaq Singh Valdiya, acting Vice-Chancellor of Kumaon University and an authority on Himalayan Geology, has, on the basis of his lifelong field work and on the spot studies in Central Himalaya, analysed the causes of these mishaps. According to him, geological changes are taking place in the mountain range, specially at the conjunction points of the lower and middle Himalayas and middle and higher Himalayas.

Prof. Valdiya says that the scenes of major disasters are located in these regions and suggests the elimination of human interference. The excessive use of high explosives for construction and repairs of roads destabilises the fragile hills. During the past three decades since the Border Roads Organisation has taken up the development of hill roads, the use of explosives has crossed all limits. High explosives are also used liberally in power projects. The sinking of the Kamak village during the October 20 tremor was due to this disastrous activity, which hollowed out the site during the construction of the Meneri-Bali project tunnel.

Similarly, Tehri town is constantly under the threat of artificial tremors caused by the use of explosives at the dam sites. The whole hill side has been shaken and houses have developed cracks.

Deforestation is another cause for the destabilization of the slopes. It is not only felling of trees, but also the change in forest species that affects the natural environment. Since the introduction of commerical forestry, mixed forests have been replaced with monoculture of coniferous stands, which are lucrative timber mines. Coniferous trees do not encourage undergrowth.

Though a moratorium on the felling of green trees for commercial purposes was put in April 1981, surreptitious felling continues at an indiscriminate rate. This was with the connivance of the state government as it provides work to state government undertakings like the UP Forest Corporation and increases revenue. Moreover, the timber mafia who enjoy the political patronage of all parties, have organised the mass felling of trees under the pretext of purchase from private owners.

The alpine pastures, which form a protective cover between the tree-line and the snow line are also being ravaged by the buffaloes of the nomadic Gujjars, who live in these areas above altitudes of 3,000 m during summer. Nature provides a protective cover of smooth grass to the soft soil which after being trampled by the hoofs of the buffaloes, creates deep scars in this fragile region. The flow of rain water and snowmelt along with soil erosion is thus accelerated.

The origin of the 1978 Dabrani landslide was in Dayara alpine pasture at an altitude of about 4,000 m above sea level. The local population in these sparsely populated areas rear only cows, horses, sheep and goats. They objected to the entry of buffaloes in this fragile region but the Forest Department's patronage overcame these odds. Government machinery works very smoothly when greased with it.

The Uttarkashi earthquake on October 20 created new threats. Fissures and cracks have developed on the hill tops. As each subsequent rainy season advances, these cracks are filled with water, the soil cover loosens and the possibility of landslides increases.

After the Uttarkashi earthquake, the first demand of the people was to conduct a survey of the danger spots and availability of correct information about the intensity of tremors. While scientists have forecast the probability of an earthquake of a higher intensity in this region, the authorities are silent. Instead, more hydropower projects are proposed and are being erected.

The logic behind this thinking is that everything in Nature—water, timber, minerals and even scenic beauty—should be converted into cash. That is what the religion of economics teaches. Conversion into cash requires a good market, which of course, is readily available in big cities.

The high priests of development in Delhi and Lucknow eulogise development prospects as symbols of prosperity. If the dangers of these projects, which are always kept secret from the people happen to get highlighted by persons concerned about safety and ecology these individuals are condemned as anti-development and therefore anti-national.

Thus, in the Himalayas both the people and the land continue to face the fury of nature and the apathy of insensitive, short-sighted and greedy government policies. Those living in the Himalayas have to struggle for survival as well as for the survival of this great mountain-system on which depends the prosperity of a vast majority of people living in the foothills and flood-plains of the Himalayan rivers.

Those engaged in this struggle have come out with a charter which has analysed the situation in these words: "We are facing a crisis of survival due to the suicidal activities being carried out in the name of development. These are commercial exploitation of forests, mining, construction of dams which inundate vast areas and displace large number of people, mechanical heavy explosions creating artificial tremors, construction of five-star hotels for luxury tourism and multi-storeyed buildings..... The Uttarkashi earthquake was a warning to stop tampering with the Himalayas which - fell into" deaf ears.

There is a need of new long-term policy for the protection of the dying Himalayas, whose main objective should be to heal the wounds of the scarred mountain ranges and bring about the economic development of the people by using natural resources in such a way that future generations are benefitted in a sustained manner without causing any harm to the fragile ecosystem.

3

Shri Sunder Lal Bahuguna's Book Inauguration

I recall an interesting meeting with Mr. Bahuguna during May 1999. He gave me a ring and invited me at the occasion of the inauguration of his book *"Fire in the Heart, Fire wood on the Back"* on 20th May 1997 by Dr. Karan Singh at Sannidhi Ashram near Rajghat. It was a nice occasion. Mr. S.L. Bahuguna met me at the gate and after a brief moment said that he desired to arrange the function under a tree but mosquitoes problem was there. We talked intimately for a few minutes.

Dr. Karan Singh arrived exactly at 6 PM and the function started with very few people, mostly journalists, from press. As time passed the speaker gave mike to the President Dr. V.N. Narayan, Editor *The then Hindustan Times*; Dr. Karan Singh; Mr. M.P. Chaturvedi, MP, Mr. Yashpal Jain and Mr. S.L. Bahugunaji.

The subject matter was Himalayas and environmental protection. Dr. Karan Singh said if Himalayas will not stay, then it would be impossible for the country to sustain. This povides all rivers and water resources. Even the world mountains Andes, Alps, and Aravalis are sources of water.

The book was highly praised and Mr. Bahuguna's efforts

were appreciated. Mr. Bahuguna said that in my will, I leave to younger generation the job of protecting Himalayas as he is old and is feeling difficult to move the authorities though, the latter has made so many promises to him already.

There was a brief Uttranchal movement (before Uttranchal declaration) as to why, Dr. Karan Singh said, the promises were not being kept and would that be only done when there would be bloodshed? Why not before that.

Dr. Narayanan said he has been publishing Mr. Bahuguna's articles in which one is included in the book. He said he is a fan of Himalaya's and desires its protection.

Dr. Karan Singh said that he belonged to Uttrakhand. His mother was from Himachal and he remained in touch with West Himalaya's throughout. He said Himalayas need all sort of due protection.

The function ended after an hour's duration with vote of thanks. It was a simple yet very attractive occasion.

4

Silent Fast Day of Shri Sunder Lal Bahuguna

Mr. Bahuguna gave me a ring from Tehri that he would be available on 18.1.1997 at 1 p.m. at Sannidhi Ashram, Delhi. Also one R.D. Sharma, his relative, was asked to be given a ring to contact him. Unfortunately the slip remained with me and on 18th only I saw that at 1 p.m. and then speedily rushed to a phone and rang up. He said he was not available earlier which gave a relief. Any how we met him that day. I was to give an interview on 20th. Earlier I had sent him a New Year greeting alongwith a letter of congratulations for getting Best Citizen Award.

On 18th Jan. when I met him at Sannidhi he was ready to move to Chennai for attending a symposium. He was on mook vrata (silent fast) and wrote lines and lines to express himself. He was so active that I could not believe. Ultimately we decided to meet Mr. Jai Prakash Aggarwal, a sitting MP from Delhi from Chandni Chowk that time.

On the way near Golcha he was to meet one publisher. Also, he had prepared one floppy of 4 articles and one was remaining which he asked me to complete and give to his publisher. We met him. But I expressed my inability as just after the interview on 20th, on 21st I had to go to Dehradun for my child's admission. All the way I was worried about his

Chennai Train timings of 3.50 p.m. on 26th as already it was 2 O'clock. All along in the car he was continuously writing and talking to me with pen and paper. This was of course dangerous also as in the busy streets of Chandni Chowk but any way I enjoyed that.

At Chandni Chowk near Annapurna I parked my car at non-parking site and went to Maliwara to see Mr. Jai Parkash Aggarwal, who was not at home. A long chit we left for him. The chit contained my interview dates and a request to talk to a V.I.P. of Delhi for the same.

He desired to purchase films but said that it may take some time and that he had no money brought and eventually I too did not possess sufficient money with me and so purchasing was dropped. When we were returning suddenly a police cop, was found noting my car number in his challan file when I reached. I requested the cop that it was a staff car and on emergency that was parked. So he allowed us to go.

On the way he told that J.P. Aggarwal belongs to Srinagar Garhwal and is more sympathetic with Garhwalis. He gave also reference to BJP MLA Mr. Panwar. I told him that I know him well. Already I had secured an appreciation letter from Shri Bahugunaji which I said I would produce in the interview. He had written a few lines on that for me.

While returning he asked as to where my daughter would stay at Doon and some references of his relatives, his brother, who lived opposite a hotel at Ballupur.

Back at Sannidhi we reached around 2.40 p.m. and met Mrs. Bahuguna who met me affectionately. The luggage was loaded in the car. Actually initially he had asked whether I would be free to leave them at railway station. Yes I said. He showed me his pass and train ticket.

On the way he wrote two letters, one to *Hindustan Times* and the other to Mr. Rao at Hyderabad. He gave me stamps and asked me to post them. Also he gave me a cheque of Rs. 1500/- to be given to publishers to contact his sender to send papers to me so that I could take them to Dehradoon as I was leaving on 21st Jan. I dropped him and his wife at Nizamuddin Railway Station. I was happy that I had learnt something new that day from him.

5

मीरा बहन का प्रकृति प्यार

राधा भट्ट

गांधी युग की महान साधिका एवं सेविका मीरा बहन का जन्म शताब्दी वर्ष मनाने के लिए हिमालय सेवा संघ ने एक ''मीरा बहन जन्म शताब्दी समिति'' का गठन किया तथा उसके तत्वावधान में 22 नवम्बर, 1992 से 22 नवम्बर, 1993 तक जन्म शताब्दी कार्यक्रमों का आयोजन किया गया। इन कार्यक्रमों का प्रारम्भ राष्ट्रपति भवन में आयोजित एक भव्य समारोह में महामहिम राष्ट्रपति डा० शंकर दयाल शर्मा ने 26 नवम्बर, 1992 को किया।

हिमालय के प्रति मीरा बहन का एक आन्तरिक आकर्षण था। उन्होंने अपना प्रथम एवं विशाल सेवा कार्य हिमालय की तलहटी में गंगा जी के किनारे पशुलोक आश्रम के रूप में प्रारम्भ किया था। यहीं एक विशाल भूखण्ड में उन्होंने गांधी जी के ग्राम स्वराज के कार्य को साकार रूप देने के लिए बापूग्राम की स्थापना की थी। इस गांव में 24-25 मार्च, 1993 में मीरा बहन जन्म शताब्दी कार्यक्रम के अन्तर्गत एक पर्यावरण चेतना शिविर का आयोजन किया गया जिसमें उत्तराखंड के कार्यकर्ता--साथी एवं बापूग्राम (ऋषिकेश) तथा उसके आसपास के गांवों के ग्रामवासियों ने भाग लिया। इसी प्रकार का एक पर्यावरण चेतना शिविर गांधी सेवा आश्रम, ओयल, ऊना (हिमाचल प्रदेश), जहां मीरा बहन ने कुछ समय तक कार्य किया था, में भी लगाया गया।

जन्म शताब्दी वर्ष के अन्तर्गत एक महत्वपूर्ण कार्य मीरा बहन के जन्म शताब्दी ग्रंथों का प्रकाशन था। हिन्दी व अंग्रेजी में प्रकाशित इन दो ग्रंथों

में मीरा बहन की जीवनी तथा उनके चुने हुए लेखों में साथ बापू के, उनके नाम कुछ विशेष पत्रों का समावेश किया गया है। साथ ही वे गिने-चुने संस्मरण भी इन ग्रंथों में संकलित हैं, जो मीरा बहन को निकट से जानने वालों ने उनके तेजस्वी व्यक्तित्व के विभिन्न पहलुओं को उजागर करते हुए लिखे हैं।

हिन्दी व अंग्रेजी के इन जन्म शताब्दी ग्रंथों के अतिरिक्त बापू के छः सौ से भी अधिक पत्रों में से, जो कि उन्होंने मीरा बहन को लिखे थे, एक सौ ऐसे अंश एकत्र किये गये हैं जोकि हर सत्य शोधक के लिए प्रेरणा व चिन्तन की सामग्री है। इस पुस्तक का नाम है ''ड्रोप्स ऑफ ड्यु'' (Drops of Dew)। यह सचमुच ''स्वर्गीय ओस बिन्दु'' के मोती ही हैं।

ये पुस्तकें हिमालय सेवा संघ, राजघाट, नई दिल्ली-110002 से प्राप्त की जा सकती हैं। हिमालय सेवा संघ ने अन्य अनेक पत्रिकाओं में मीरा बहन के बारे में लेख भी लिखे तथा मीरा बहन के बारे में विशेषांक भी निकलवाये।

हिमालय व वहां के जन-जीवन के बारे में मीरा बहन का जो विशेष लगाव था उससे उनकी कर्मभूमि मुख्यतः टिहरी जिला बन गया था। साथ ही हिमालय के विकास को सही दिशा देने का अथक परिश्रम भी उन्होंने यहीं किया था। उन्होंने श्रम, स्वावलम्बन व प्रकृति के साथ समरसता रखने वाली यहां की जीवन शैली की शक्ति को अधिक सक्षम बनाने के लिए गांधी जी के ग्राम स्वराज्य के अनुरूप उसे आगे बढ़ाने का प्रयास किया था। हिमालय के पर्यावरण ह्रास के प्रति उस समय भी उन्होंने पूरे देश को चेताया था 'Something is Wrong in Himalayas' जैसे अपने लेखों को लिख कर। भारत भूमि का विकास किस तरह शुद्ध भारतीय हो, इसकी उन्होंने बहुत कोशिश की थी। हिमालय को बांज वृक्षों के आच्छादन से पूरी तरह ढकने से ही हिमालय का अस्तित्व कायम रहेगा, ऐसी दूरदर्शितापूर्ण गुहार उन्होंने लगाई थी। उनके इन्हीं संदेशों को उनकी जन्म शताब्दी के अवसर पर आज के भोगवादी विकास के लहूलुहान हिमालय के जन समाज तक पहुंचाने के लिए एक 14-दिवसीय पदयात्रा का आयोजन 30 अक्टूबर से 12 नवम्बर तक ''गोपाल आश्रम'' गेंउली, टिहरी गढ़वाल से बापूग्राम, वीरभद्र, ऋषिकेश तक किया गया था। इस पदयात्रा टोली में प्रतिदिन 20 से अधिक पदयात्री शामिल रहे, जिसमें तीन-चौथाई से भी अधिक संख्या बहनों की थी।

225 किलोमीटर की इस यात्रा में 17 गांवों तथा 12 विद्यालयों में सभायें कीं, साहित्य व संदेश तो पूरी राह वितरित किया गया।

व्यापारिक दोहन के अन्तर्गत कटते वन, दैत्याकार बांधों के नीचे डूबती पहाड़ी घाटियों की एकमात्र समतल उपजाऊ भूमि तथा पांच सितारा पर्यटन जैसे उद्योगों के द्वारा प्रदूषित होते पर्वतीय जनसमाज के गम्भीर संकटों से इस टोली ने लोगों को सावधान किया तथा हिमालय एवं हिमालय के द्वारा पूरे देश को बचाने के लिए पहाड़ी ढलानों को हरियाली से ढकने व पहाड़ी कृषि को ग्रामोद्योगों द्वारा पुष्ट करने के लिए लघु जल विद्युत के प्रयास हर पहाड़ी नाले-झरने पर करने का संकल्प भी किया।

गेंउली (गोपाल आश्रम) में ग्रामवासियों ने एक मीरा बाल शिक्षा निकेतन का काम प्रारम्भ किया है। अपने क्षेत्र के विकास को मीरा बहन द्वारा दी गई दिशा में बढ़ाने के लिए वे उस क्षेत्र में जड़ी-बूटी की खेती का काम विकसित करना चाहते हैं।

इसी तरह बापूग्राम में मीरा बहन के कुटीर को केन्द्र बना कर वहां के लोग महिलाओं व पुरूषों की सामूहिक स्वावलम्बी शक्ति गठित करने की ओर विचार कर रहे हैं।

भिलंगना घाटी में गेंउली के आसपास के कई गांव बांज वृक्षों का रोपण कर मीरा स्मृति वनों को बनाना चाहते हैं।

इन सारे प्रयासों को प्रोत्साहित करते रहने का काम हिमालय सेवा संघ के सामने है।

पूर्वोत्तर एवं पश्चिमोत्तर हिमालय के सभी कार्यकर्ता साथियों की एक गोष्ठी फरवरी में मीरा बहन जन्म शताब्दी आयोजनों के अन्तर्गत ही आयोजित करने का निर्णय लिया गया है।

इस प्रकार उस महान आत्मा को हिमालय सेवा संघ ने ''पत्र पुष्प'' के रूप में अपनी नम्र श्रद्धांजलि अर्पित की है। इसमें सबसे महत्वपूर्ण भावना एवं समर्पण, योगदान व श्रम हिमालय सेवा संघ के संस्थापक एवं भूतपूर्व मंत्री श्री कृष्णमूर्ति गुप्ता जी का रहा है। उनकी मीरा बहन के प्रति जो अटूट श्रद्धा थी उसकी अभिव्यक्ति उन्होंने इस कार्यक्रम के द्वारा की है।

6

Creating Environmental Awareness

The environmental awareness which persisted since long from the beginning of the history, among almost all human settlements, have left behind exemplary traces—Babylon's hanging gardens, the green avenues of Versailles etc. that reveal the highest degree of the sense of discipline, sensitivity and regard which man had shown towards nature. Whether it was a one man's show or a joint effort of community, all activities were oriented to see things beautiful. A deep sanctity prevailed among masses for the natural features to preserve them in the most natural and purest form which even to date are being followed, though, not so vigorously.

The planning of city of Mohenjodaro 3000 B.C. along Indus river is another example where a very high degree of the planning principles were adopted. Among beautiful man-made environment, came the Inan fortress usually called Inca, capital of Sacsahuman clinging to a hill top north of Cuzco, built somewhere in 14th century. Similarly some of the left out majestic raws of mammoth on Easter Island 12th to 17th centuries trace out the rich artistic taste of the civilization of that period. The beautiful temple of Parthenon in the city of Pericles still towers over Athens from the top of Acropolis that perfectly blends man and nature. The ruins at Delhi couched in their awesome suroundings are no less interesting. Similarly are the Persepolis ruins that still look elegant and majestic. Taj Mahal at Agra is a unique example in the history of monuments. The monuments at Kremlin too have no match.

Several artists with their brush strokes described how beautiful the nature is or *vice versa*. Braqque's, 'Bird returning to the nest', Van Gogh's 'the Cypresses', Raphael's 'The School of Athens', Picasso's 'Cubism', and works of other artists—Mikhail Ani Kushin, Alexei Gritsai, Svetoslav Roerish, Tatyana Shimes, Amrita Shergil, Hussain are marvellous. Various poets, writers, dramatists too contributed much in bringing out landscape visions in their works. Words-worth's 'Lucy Grey', Robert Southley's 'Inch Cape Rock', and similarly work of other poets like Shelly, Keats, Bayron and other's is worth mentioning. Daniel Defoe's 'Robinson Crusoe' shows deep love of the author which he had for the sea and similarly Johanna Spyri's 'Heidi' is one of the most interesting writing on nature, which describes about natural elements—trees, goat, butter, monuments, icy hills and love of little children with old people.

Suddenly the degradation of environment started when the human settlements felt the vagaries of nature unbearable. The draughts, famines, and other climatic disasters, though, they persisted earlier too and tolled heavy lives, gave masses a new direction to think for self protection. This started the industrial revolution all over the globe and a large scale migration took place. Industrial slums were formed and so grew up the concept of towns and cities. Intensive industrialisation gave birth to the word pollution and there was then no end of the latter as the greed of the entrepreneurs multiplied. The epidemics broke out all over. The filth and wastes got accumulated on roads. The quality of life became very much degraded.

The root cause of the environmental degradation was explosion of population as large scale industrialisation and commercialisation took place which swallowed away the major green areas. These congestions and over crowdings, compelled people to squat all over which thereby spoiled all natural features of the cities. The sea, river, lakes became the large sites for dumping of wastes which created severe water pollution. The vehicular traffic created major air pollution. The flora and fauna gradually started disappearing. Heavy deforestation started, which adversely effected the climatic conditions.

For a long time the system moved like that only until intellectuals and environmentalists felt the adverse pollution and degrading ecological imbalances, against which they raised their voices which, however, did not receive appreciation from the greedy quarters. However, a movement took place subsequently and global alertness was aroused by which the various country heads started taking interest as they showed concern over such pollution effects and felt the dire need of conservation. CHIPKO movement from India, 'Save Buna Movement' of Japan and 'Pahar Bachao Andolan' (PBA) from India for saving ridges and hills came in action which subsequently became global slogans.

The large scale cultivation of forest land, the slash and burning of trees, the forest fire and the poachers activities created much damage to the forests. No doubt, the livelihood of a large number of people depended on poachers and cultivation, even then the few governments like that of Japan, Thailand, Zimbabwe, Kenya became tough against them, to safeguard their forests, as they realised that its denundation may create draughts, floods, famines and all types of such adversities. The forest needs were more felt as it was also realised that several plant species were dying and recycling process dwindled as the seeds no more remained protected.

The increase of population has been held responsible for dwindling away of the rare animal species and if it remains so feel Africans, the day is not far when the animal life is bound for extinction. As a conservation measure the hunting has been banned at several places. The heavy casualty of rhinoceros, elephants and other wild animals in Africa on account of the greed of poachers for rhinoceros horns, elephant trunks have created a great alertness among the African authorities particularly in Zimbabwe, as even 'shoot at site' orders have been flanked up for those indulging in such activities irrespective of their being natives or outsiders.

However, at places like America where it is observed that certain species of population—alligators, leopards was increasing so ban on hunting was lifted. The hunting, it was realised, indirectly checked the poachers activities. In Delhi it is very sad to know that most of the ridge fauna disappeared on account of quarrying operations and large scale urbanisation.

At places, like in Japan, certain restrictions have been put up on fishing of some rare species but poachers are busy using mercury salts to achieve their goals. Similarly for Dolphins America has restricted for a maximum catch of 25,000 a year. In the Victoria Lake of Africa, such precautions are being taken to save haplocromines variety of local fishes which are fast dwindling.

The projects for conservations of the birds have now been taken up in various countries like China, America, Japan, India and others as a result of which several bird sanctuaries have been developed. Late Dr. Salim Ali, a cute bird watcher from India did tremendous work by taking up hard expeditions in remote areas of the various countries, to know and then tell the masses about them in terms of their protection. The draughts conditions, however, made such sanctuaries failure which as such is a matter of great concern.

The air pollution caused by the industries and vehicular traffic is receiving world-wide attention and various researches—administratively and scientifically are being done to check the effect of foul gases—Sulphur dioxide (SO_2), Hydrogen sulphide (H_2S), Hydrogen peroxide (H_2O_2) which are damaging the ecospheres and acid rains that pollute the land, water and every thing. Moreover the balance of ozon and carbon dioxide present in atmosphere which remain disturbed and which could be harmful to the bio life, is seriously being attended. Ozon's depletion could cause skin cancer whereas increase of carbon dioxide may swell up the sea water which thereby can spoil all beaches and the various coastal settlements. The smoke of vehicles particular in cold countries which give rise to serious eye troubles and other ailments like Asthma, etc., are being seriously paid attention. In Switzerland as a measure to reduce smoke pollution the car movements are being restricted, as ban on plying of vehicles on Sundays have been introduced.

The noise is one of the serious pollution evil which could even cause deafness. Even a child in mother's womb is not safe from the noise outside as the whooping sound may lead her to death. Now-a-days several countries are enforcing codes to restrict the sound level as well as timings for convenience of the public.

The water pollution is responsible for causing epidemics and various virus fevers. The dumping of wastes in sea has become the common phenomenon as the merchant ships empty up the vessels over there only. The garbage which goes down immediately remains unchurned for months as there is very little motion beyond a certain depth in sea, so with the pollutants, several sea animal casualties usually occur. The beautiful sea beaches are getting spoiled as the garbage floats over to the shores. Identically rivers are getting the city sewage and are becoming big sewers. The undeveloped river fronts attract all illegal constructions, industries which pollute the river like any thing. River Rhine's pollution by Sandoz factory in Switzerland had created havoc in the area which killed several men and fishes. Now various anti-pollution projects have started like Central Ganga Authority in India which is looking after the pollution of river Ganges. Similarly various codes—national and international have been framed up for banning the dumping of wastes in sea or in rivers.

In California the wastes are being incinerated as per the latest technology. But in developing countries these dumping sites are the eye sores as huge mountains of it could be found lying near river or else along busy roads. Of course recycling of the garbage is now being advocated for building construction purposes by mixing it with fly ash and converting it into bricks but so far it is not being totally accepted.

The continuation of thousand years old mountain ranges now remain broken by constructing roads over there and developing the areas. Delhi experiences a number of such rock casualties all over the city where at one time had flourished very interesting flora and fauna which have now all disappeared due to continuous blasting and quarrying operations. The creation of New Delhi on Raisina hills was perhaps the first encroachment, over ridge and then followed so many.

The grace, charm and grandeur that earlier open spaces had, are not seen now, as it is believed that, it were our predecessors only who took keen interest in such developments. Even in Delhi wherever the open spaces are available are of predated or of Mughal's time. The open spaces, within a reach of a common man are meaningful to the settlement but open space reservoirs are also essentially required for keeping the overall ecological balance. These are, it is found, fast disappearing. Merely keeping open without any plantations or life and leaving it as deserted, also does not mean anything. The Delhi DIZ area in the past had enormous play spaces for the children, which in the present set up are completely missing. So the children are becoming more indoor oriented. They read comics rather then mixing and becoming social via media of play spaces. Now no one is found playing hockey, football or such games as no park is so big to accommodate these game pitches.

Monuments, it is now being realised, are important for the cities. The enormous open spaces which they contained inside and outside are the only effective lungs remaining now, for our busy cities. So in order to save futuristic cities, creation of such buildings are essential in some form or the other as these would be indirect measures for saving our lungs. Otherwise the greed of land would swallow away all meaningful open spaces.

The old congested settlements like walled city which have become commercial centres are getting degraded due to its congestion and traffic problems. Though, they are still graceful and the inhabitants enjoy living over there, even then, such areas require a conservative surgery. Adequate parking spaces have to be provided over there and its major lungs require adequate protection.

As city authorities are not able to provide housing to all and within their budgets, illegal squatting on land has become a universal phenomenon. The life is so degrading over there, that even the desired facilities—drinking water, electricity, roads, sewers are not available. Even proper schools and medical facilities do not exist. It is only when the authorities feel interested that subsidised standards are formulated for them and very slowly they are provided over there and this process even takes as much as 50 years time. The urban villages too face such problems.

Most of the villages look like perfect urban areas where intensive construction work is in progress. What these people require is a kind of mixed living pattern which should be both rural as well as urban.

The housing in cities for weaker sections is far from being satisfactory. Even 12 square yard parameters have now been suggested by the authorities for a person who lives in a joint family that comprises of two or three families. What environmental decoration could anybody expect from such quarters, in the present time, when we are heading for 21st century? We are still not providing attached bath and latrines to individual homes and are depending, as such on community taps and toilets. Still people go in open for ablution purposes, meaning thereby that housing requires lot of improvement. Moreover, houses are expensive and are beyond the affordable limit of the major population of the cities. Along with this, there are other planning problems too like houses having no verandahs, no sun light, no ventilation etc. which make the environment very dull.

The overall city system is facing a number of problems. All the building bye-laws, development controls and such measures are though being enforced but the public wants to evade all these. Of course, such measures are helpful in bringing out a uniform sky line. The lack of provision of bulk services in advance—telephone cables, sewers, drinking water, roads, causes continuous digging of roads which is very expensive and awesome. Moreover, in an expanded city, going over to central places to enjoy certain facility, how much money and time one has to spend. Can't these facilities be dispersed? The role of the architects, planners, critics, journalists is found to be immense, in developing a healthy city, which should be honoured.

7

The Moghul Palace—Red Fort, Shahajahanabad

["The majestic Red Fort needed sympathetic but strong treatment. The real problem was to reconstruct the landscape and recreate the old environment. It could only be possible if the Army was moved out of there first. The Red Fort is one of the difficult places to be conserved. It would need a special task force of individuals who understand conservation and how exactly to conserve it retaining much of the old character."

—Dr. Bernard M. Fieldon.]

Anybody whosoever may go to visit Red Fort at Delhi, it would be difficult for him to locate it, as the site that happened to be visible from far and wide with beautiful green lawns, is totally now eclipsed with parking and odd traffic. At least, if not we, than the tourists, who come to see them from far off places, mind it. They don't pay for visiting our parking lots. Moreover inside the Fort, there is very little to see, as almost all surroundings are still occupied with the cantonment area comprising of tall barracks, horse stables, residences of army personnel and so on. For the tourist it is not possible to judge the actual grandeur of the monument, as they can not go over to the places occupied by the Army. How sad it is, that even years after Britishers have left this country, after spoiling our heritage, we have not yet been able to restore that, despite there being all facilities available to us. After all what those barracks are doing over there, which are spoiling the grand skyline of the complex as seen from the river side. On top of this, ironically a Ring Road has been built on the east side, just touching the palace which has spoiled the grand silhouette of the monument. Towards north of the Fort, a huge colony has come up amidst the heavy traffic. The major road in front of Red Fort, carries traffic which disturbs the sensitive environs of the area. At the palace where the queen's gardens and other parks were there; are now situated, Lajpat Rai Market and other such complexes. Meaning thereby that a huge mess has been created all-round the fort.

Red Fort is a large citadel and a copy of the Agra Fort, the latter, though, being much better than the former in all respects. Its public entrance is now through Lahori Gate that lies on the axis that passes through Chandni Chowk. It is built with red sand stone which looks distinct from far off areas. A moat about 10 m wide, that runs allround, was put up as a barrier against any enemy attack. The Jama Masjid, situated towards South West of it, is a sort of complimentary structure to it. The Salimgarh Fort is an important building toward north of it which was joined through an arch bridge, traces of which still could be seen. The back side of the Fort contained beautiful palaces facing the River Yamuna. From there was visible Shahdara a small township, across the river which happened to be a *mandi* during those early days.

Shahajahan after a reign of 11 years at Agra, resolved to transfer his capital to Delhi. He ordered his engineers to prepare a plan for a palace similar to that of Agra and Lahore. Its foundation stone was laid in 1638 AD. It was completed in 1648 AD. Izzat Khan, Alah Vardi Khan, Makramat Khan, Ahmad and Hamid are the names associated with it as its architects and engineers.

Perhaps every stone that was fixed on the walls, floors

and roof must have been approved by the Emperor himself, before laying, as it appears from the workmanship which is quite appreciable. The artisan role in building the total complex had been supreme as even under pressure and without even getting salaries they worked.

Through Lahori Gate, one finds a large Gateway leading to a courtyard, enclosed on three sides, and a way further going to the royal *bazar* of those old days, glimpse of which still could be seen today. Today even the very aristocratic shops are doing brisk business of antiques over there. The roof structure is of very fine dome with arches all over. For this area it is said that on present upper floor level, the traders from as far as from Persia, Gandhar and Agra used to sell jewellery, gems, clothes and other such fine expensive items. Further towards east side is situated a Gate, on the upper floor of which is now functioning a Museum, where arms and weapons of the past, have been displayed. From there is visible the famous Diwan-I-Aam, which is reached after crossing a big lawn. Diwan-I-Aam is a large hall of pillars, with the famous Peacock throne, on one side of it. From there one could go over to the Palace Complex.

We get a number of serial visions as we go further. The palace, mosque, sheeshmahal are the other buildings which are grouped together, facing eastern side, towards Yamuna. In one of the halls, runs a museum, where are displayed, the costumes and rare books related to Moghul period. The gardens are laid beautifully with fountains. The rose water earlier flowed through the ladies palaces the latter having fine marble jally work allround. The architectural details of the carved pillars, arches, domes, marble flooring was exclusive of the era. Even roofs contained gold plated work, which the invaders from time to time looted after melting that.

The Moghuls gave all due respect and comforts to their ladies as they were busy mostly in fighting wars or attending other important jobs. The Palace, as seen from the Yamuna river side looks very beautiful. Its white marble domes, red stone wall, arches, jally work has no match with the architecture of any time in the history. The Gates, through which they entered the river, to boat or to swim could still be seen.

Unfortunately now many domes have fallen off, the plaster too has peeled off and stones have given way from various places which as such, is a matter of great concern. Moreover, the stone barracks built by the English, stand too high, than, those fine structures, which look very odd.

Red Fort in the past had a continuous link with Humayun's Tomb and other monuments at Mehroli and other places and so there must be running, a frequent traffic, to those places in the early days.

A number of tourists visit the Fort every year. For their entertainment a special show called Sonet-Lumiere, which is a play of light and sound, is being run in the evening. It draws heavy crowds.

A master plan for this monument is earnestly required to be prepared by the Archaeological department. Its improvements could be made in phases. They could be filled with water again and could be switched on for other entertainment uses. At last the parking from the area be removed and let the Lajpat Rai Market area be again turned green and the market either squeezed into a multistorey building or be shifted to an alternative area.

8

Forests

The Delhi-citizens perhaps must not have forgotten that during Asiad 85, despite strong opposition from the ecologists, the journalists and the environmentalists how a number of matured neem trees were butchered from Lodi Road area and the road was ultimately widened. There was no effort for fresh plantation in place of the deseased ones, as say, after, a fortnight of Asia programme it was a be-gone and a by-gone story for which none came forward to shoulder the responsibility of dressing environmental wounds or to acknowledge the assurances given to the public earlier, before axing, that they are conscious about not felling of the matured trees and that if it was so they would take adequate measures. But any way, was it really just to axe down age old trees just for the sake of a week long function despite of its being prestigious programme for India. And similar such operations were repeated at Tughlakkabad and in other parts of the city.

We agree that the above instance took place in hasty measure and that too in an atmosphere of known consequences. But what about such massacres which are daily taking place despite the laws and rules being tough for the offenders. Then could Delhi have a better future in terms of environment?

The euclyptus war has invaded the total capital's environment from which there is truely no escape. No doubt, it has given good results by getting the forests developed in no time but ultimately at what cost? Do we think that Delhi's semi-desert conditions could any more sustain such imported trees which thrive on water only? Is it not a fact that their growth has lowered down sufficiently the subsoil water level in the capital? Is it not ironicall that their fast growth and easy availability tempted one to plant them in place of wild natural growth which earlier had dominated the entire city? In groups they may look all right but what about while they are grown in isolation when they give a shabby look. All along Yamuna and in the trans-Yamuna areas large euclyptus plantation has been done. Similarly along Western Yamuna Capital tall grasses called 'Bhara' have been removed and endyptus trees have been planted. How was that the early planters were not conscious of its after measures? All right, let's admit our mistake but are we now thinking to correct it by replacing all those trees with fresh plantation of indigenous trees or just would sit down with closed eyes.

The Mughals tastefully planted palm trees in their parks which to date even could be seen at Subhash Park, Qudsia Gardens and at such other places. However, with the time some trees have met with casuality. Is it not surprising that after Mughals nowhere such examples were repeated which as such clearly signifies that the later generation did not have at all any taste, liking or craze for enhancement of the environment? Only one example, however, we do come across, that is near Pitampura, along western Yamuna canal, where under Sh. M.N. Buch a Salwood forest was developed as an experiment and where trees now have grown to about five metres height in about eight years period.

Similarly along Chandni Chowk where there stood trees along both sides of the road; none exist now. In the city area too, the business concentrations removed all such elements.

The Red Fort lawns remained almost without trees; meaning thereby, that in Delhi, if there was any area, where maximum harm was done to the environment, it was Shahjahanabad.

No doubt for making Delhi as the garden city of India no less efforts were made. Sir Edwin Lutyens with the help of Sir Walter George got several varieties of indigenous trees planted along Janpath and in other parts of New Delhi areas. The irony, however, was that along Shahid Bhagat Singh Marg, by the late sixty's almost all matured *'Imlee'* trees vanished. Similarly all *Dhaturu* trees growing at Baird Road, Panchquian Road and at other places along road sides as well as in parks disappeared. The heavy tarcoaling on roads is also one of the factors responsible for uprooting of the trees. The Shisham trees which were planted in parks and on road sides too fell down either due to strong wind pressure or on account of heavy rains, meaning thereby that Delhi's plants met a cruel treatment either from the nature or from the man.

Lutyens planted *Jamun* trees along Rajpath lawns and in DIZ area and on the road sides of New Delhi. Along Ashoka Road, Parliament Street and at other places still one could see huge indigenous trees grown up there. At Talkatora Gardens a number of mango trees grew up. The *amaltas* trees too are found on road sides and in the parks. Some of the streets like Sunehri Bagh Road, came up with beautiful get up, with these trees.

The Delhi ridge remained the only forest belt of the capital where mostly *kikar* trees had grown. The Master Plan of Delhi published in 1962 too mentioned that ridge forest should be protected and preserved. But as the ridge was invaded from various sides, the forest declined considerably. Otherwise the ridge contained at one time a rich flora that constituted trees like *kikar, katha, phulahi, neem, dhak, pilu, hingot, dhoy, lasoona, dhak, haitho* etc. A large number of herbs and other vegetarian plants too existed in the area important of which were *lapa, missi, jhijhri, punarnava, hulhul, gokhru* etc

The rich wild life over the ridge was worst affected due to continuous blastings and quarrying operations. The Kushak nala and other such areas where varieties of trees, birds existed are still being cruelly handled as trees are being axed down over there.

The disappearance of wild life on account of fast urbanisation, indirectly finished the forest areas. Earlier to feed the domestic animals one felt the dire need of their food which were leaves of mulberry, *pipple* and such other trees. In small colonies people do not like tree plantations as that create hindrance for allowing sunlight and more over every one prefers more paved spaces.

More nurseries are required from where even matured trees could be taken for development of forests. Several experiments and researches should be worked out in this direction. The orchards should be protected and preserved and more should be developed to meet the fruit requirements of the city. The forestry is desired to be the peoples movement. The public may be given incentive to keep pet animals like goats, sheep, cows, horses so as to indirectly help the growth of trees. Near residential colonies exclusive areas be marked for dairies.

The old matured trees required replacement by new. The seeds must be well protected. In the city, the botanical gardens could be planned for which the possible site could be along western Yamuna canal. This could be an instrument to boost up the plant life. The cutting and felling of trees be completely banned inside the city. In district parks and other big areas extensive afforestation programme be taken up. The building of roads where trees exist should be banned. The deforestation could otherwise manifest in numerous ways like floods, land slides, soil erosion, silting of canals and reservoirs, etc.

9

Aravali Ranges

Linking of Two Ridges

The Delhi's ridge—a range of Aravali mountains—which in the past extended from Dhaula Kuan and beyond in the South-West to the river Yamuna in the north-east was unfortunately broken into two parts: the south ridge and the north ridge by a large number of human settlements which agglomerated in the areas of Karol Bagh, Motia Khan, Filmistan, Pulbangash and other areas of North Delhi. The north ridge was further sub-divided in the north-east side and a part of it was left behind along the river Yamuna across the present Alipur road.

Whereas all settlements grew into a typical city—like style, certain exceptional areas like Idgah remained somehow protected. The existence of the rocks and matured *kikkar* trees in the area barely testify to the presence of the ridge earlier at that very place.

At Dhaula Kuan a chain of multi-storeyed hotels have been built in addition to the existing army hutments. At Naraina the Naraina village has exploded with the growth of unauthorised constructions. At New Rajendra Nagar several encroachments have come up just along the profile of the ridge. In the Mandir Marg area and along Link Road several petrol filling stations and multi-storeyed buildings have come up. The most ironical planning proposal of the time was the construction of the Talkatora Stadium on the ridge, which has done an irreparable damage to the ecology of Delhi. Similarly developments were repeated on the north ridge side.

It was perhaps when the Delhi University was founded that the north ridge was cut off and roads were built across it for providing accesses to the University. Meanwhile, the surrounding areas of Timarpur Camp, Kamla Nagar and other areas developed and the detached part of the ridge lying across the present. Alipur Road remained very much neglected and unused. With the time two jhuggi-jhopri colonies, namely, Aruna Nagar and Old Chandrawal have come up there which have expanded on their own momentum. Moreover, beautiful *kikar* trees which were growing in the area have vanished and the soil was badly damaged due to excessive quarrying operations and on account of the dumping of waste over there.

It was unfortunate that the first Master Plan of Delhi, 1961-81 failed in providing adequate protection to the ridge as the restoration of the ridge was not planned that time.

Let us first think to restore the ridge, in terms of its continatuion first, by blending the south ridge with the north ridge. Next to it we could think of soldering the detached part of the ridge lying across the Alipur road side to the north ridge, so that even up to the trans-Yamuna area, the entire scene may look as one single landscape. For chalking out a project of this type, of course, a large amount of finance and strong legal backing would be required as the structures thus affected would definitely create legal complications. And, above all, there would be political pressures to defer the scheme.

There is only one way out, and that is like in the past, the Government had invited sensitive European architects, namely, Edwin Lutyens and Le-Carbusier for building cities of New Delhi and Chandigarh in India; similarly outstanding personalities of international repute may be called upon to work on such projects.

The Aravalis at Tughlakabad

To the Aravali hills around Tughlakabad—extending in the south-east to Suraj Kund and Haryana in the west to Mehrauli and Palam and in the north to Kalkaji-Okhla side—much damage has been done by the creation of the Mehrauli-Badarpur Road, as a large number of unauthorised structures have come up all along it at places like Khanpur, Saket, Lal Kuan, Prahladpur. They have completely destroyed the ridge and have blocked the vision of the beautiful ravines of the Aravalis, which are in lower elevation of about 5 metres high that side.

Moreover, the Mehrauli-Badarpur Road has broken the continuation of the beautiful ancient forts of Tughlakabad and Adilabad, built on these hills and have altogether separated them. An ancient over-bridge, which in the past had connected Tughlakabad to Ghiyasuddin Tughlak's tomb—later situated in an artificial lake, lies abandoned now from the centre.

The area falling on the north of Mehrauli-Badarpur Road—Chirag Delhi, Greater Kailash, Okhla—have already been extensive developed and more roads are being built over there, by cutting rocks, for connecting them with Mehrauli-Badarpur Road.

The shooting range complex has spoiled the ridge, as lawns, roads and buildings were built all over, which did not match at all with the natural surroundings. Moreover, the road that leads further in the east to Suraj Kund is bound to invite heavy traffic as well as unauthorised constructions which could be harmful to the ridge in the near future.

The strong-crushers on the south of Mehrauli-Badarpur

Road, present an ugly look and create noise and air pollution. Ahead of this site towards the east, lie unauthorised extensions of villages—Lal Kuan and Prahladpur, just along the road side, which have completely blocked the view of the ridge. The rocks in the Badarpur mines look beautiful but around it large number of labour hutments have come up.

Though it has become popular as a picnic resort, Suraj Kund has lost its natural charm. It remains heavily crowded. Grassy lawns have been laid on rocks that look unnatural and tall restaurant buildings have come up all around it from where no sky could be seen.

The approaches to Suraj Kund as well as Tughlakabad, from Mathura Road, would have been smooth and easy if the Tughlakabad railway station was not in between as the public has to wait for hours at the crossing for the green signal. Since an over-bridge is under construction, the situation may improve. Across the railway lines along Mathura Road at the Delhi-Haryana border, a number of unauthorised constructions are coming up.

The Khanpur, Lado Sarai and Saket area, situated on the west of Tughlakabad and on the south of Delhi-Mehrauli Road which were earlier small villages—have attracted a large number of unauthorised structures and have disrupted the free link which earlier existed between Tughlakabad and the Qutab. Moreover, the Sainik farm, located near to Khanpur area has been inviting building activities.

The bald rocks in the Tughlakabad belt are the richest and the safest. Efforts are required to conserve and protect them.

Traffic could be diverted to the IIT Gate from Badarpur side via Chirag Delhi and Panchsheel areas. Another proposal could be that roads may be built underground so that the ground could be made free for the pedestrians use.

All building and development activities as well as quarrying operations be banned in the area for which strict legislations are required to be framed. Quarrying operations with special permission may be allowed under certain limits. Efforts may be made to let the area remain as it is, in its natural form and colour and wild growth may be procured at all costs. The Badarpur mines area is required to be cleaned up and beautified. It may be made accessible to the common public.

The ruins of Tughlakabad require proper procurement and identification. Tourist huts could be built all over the ridge at intervals, where guides may be deputed to provide information.

Restoration of the ancient city of Pandavas, that lay between Nigam Bodh Ghat and Purana Qila and beyond to Suraj Kund is desired to be paid due attention as traces of the old faded coloured rocks belonging to that age are just found at 100 yds. away on the north of Suraj Kund.

Open spaces at Kalkaji temple, District Parks in Kalkaji, Okhla and Tughlakabad, may be blended together. If possible the Tughlakabad railway lines may be laid underground in order to properly link Mathura Road with Suraj Kund and Tughlakabad areas.

The air and noise pollution, created by the Badarpur Thermal Power Station and the stone-crushers may be minimised. Electrostatic precipitators could be used to reduce the smoke.

The continuation of the over-bridge that linked Ghiyasuddin Tughlak's tomb to Tughlakabad may be restored. Besides, the forts and the monuments of the area give adequate conservation.

Beautiful meadows could be developed all along the areas at Tughlakabad, where varieties of animals such as sheep, goats, horses, camels, zebras, deer etc. may be reared, so as to create a most natural environment.

The land allotted to the defence forces on south of Mehrauli-Badarpur may be re-examined for preservation of the rocks. The unauthorised extensions of Khanpur, Saket, Lado Sarai and other areas may be regularised and their future growth be totally banned.

The beautiful ravines of the ridge at the shooting range and at Suraj Kund side should be protected at all costs. Efforts be made to make these ridges visible by suitable planning. The wild growth may be protected.

The Save Ridge Movement (Pahar Bachao Andolan) should be popularised through mass media, to educate the public and the authorities, to protect the ridges and mountains, where there are constant threats of their destruction from the urbanisation.

The Ridge Off Qutub

The day is not far, when Qutab Minar will be encircled by tally sky-scrapers, as the ridge on the side, which belongs to the Aravali ranges, is being cut and destroyed for providing access to the various residential colonies—Palam Vihar, Vasant Kunj, Aya Nagar, Qutab Enclave, to name a few—which are fast coming up over these beautiful rocks on west side.

Along the major roads that lead to Qutab—Aurobindo Marg, Mehrauli-Badarpur Road and Palam-Mehrauli Road unauthorised extensions of the villages—Mehrauli and Lado Sarai, have come up, which have completely closed the view of the monuments.

There ridges still remain unknown to most of the people. A bird's eye view of these rocks, from Qutab, is breath-taking. One can see a vast panorama of these ridges extending in east to Tughlakabad, Suraj Kund and further to Haryana, in south to Gurgaon and in west to Palam and Rewari.

Minerals, like chalk, china-clay, mica flakes and Abrak are being extracted hardly 3 km west of Qutab in Mahipalpur and Kusumpur areas where heaps of minerals in white colour can be found deposited over the rocks.

A narrow path, which earlier lead to the villages Kishangarh and Masudpur from Qutab, has now become a major thoroughfare. It provides accesses to Vasant Kunj, Palam Vihar, Qutab Enclave, Mahipalpur, Rangpuri and to Nangal Dairy.

From Vasant Kunj to J.N.U. complex of Qutab, are beautiful ravines of the Aravalis running 1 km north to south and about 5 metres high which are in constant danger of being destroyed.

At the J.N.U. complex, which itself is located on the ridges, there are blasted rocks along the access road that further leads westwards to the residential areas of Munirka, R.K. Puram, Vasant Vihar, Anand Niketan, Dhaula Kuan and Palam.

Who ought to be blamed for the destruction of the ridges—the planners who could not foresee the fate of the ridge, the posh Delhi citizens who desired to live over there, or the authorities who gave the green signal for development proposals on these rocks?

What is done, cannot be undone now. What can, however, be done is to chalk out a strategy to save the ridge from future destruction.

Besides this, efforts may be made to restore the grandeur and beauty of the Qutab Minar which is a world renowned monument of an unique style and a symbol of our rich ancient heritage.

The ruins of the ancient site of Lalkot - the first Delhi founded by the Hindu King, Raja Anangpal and several other monuments and tombs like that of Jamali's, Kamali's, Balbans etc. which lie close to Qutab, require proper conservation and identification. Even closure of the roads be examined as they would in future invite more troubles for those rocks and the monuments. The developments taking place on west of Qutab should be discouraged and accesses to them from Qutab side be closed.

The developments should look natural and should be carried out only, where there is a dire necessity. For example, grassy lawns on the rocks sometimes look ugly, as that destroys the charm of things in their natural forms, colours and textures.

The way to Qutab from Aurobindo Marg needs to be cleaned and beautified along areas of Yusuf Sarai, Hauz Khas, NCERT Complex, Qutab Hotel and the Lado Sarai village. The unauthorised structures which have come up along the road should be removed to open the view of the Qutab from far off areas.

The surrounding villages of Mehrauli and Lado Sarai may be developed as model villages where conservation of the old village chopals, the trees, the wells and the old houses is required.

As already parks and open spaces have been created around Qutab, it could be suggested that they may properly be maintained for the growing recreational needs of the city.

Buddha Park and Rangshala

The selection of the site of New Delhi on Raisina hills was perhaps the beginning of the invasion of the Delhi ridge. It was followed by several other encroachments as a result of which the ridge lost its identity, as great confusion was created among the isolated rock fragments at places like

Qutab, Greater Kailash, Idgah, Naraina, Civil Line and Anand Parbat.

The ridge at Delhi is spoiled by the undesirable network of roads—Willingdon Crescent, Upper Ridge Road, Ganga Ram Hospital Marg, Mandir Lane, Link Road, which have been laid by destroying rocks and by felling matured *Kikar* trees of the ridge forest.

Talkatora Garden was at one time a grove of mango trees and greenery of the ridge. Unfortunately, now, it is a sight of odd lawns laid over the rocks and high boundary walls raised against the ridge. The multi-storeyed buildings of Regency Restaurant and that of Indoor Stadium, which have come up towards south and north of it, look monstrous, against the green sky line of the ridge forest.

The Willingdon Crescent, which remains busy and noisy, has disrupted the continuity of the greenery of the ridge forests; extending earleir to the Mughal Garden and rear lawns of the President's Estate. Along Sardar Patel Road, the rocks have been blasted to accommodate petrol filling stations and five-star hotels of the Maurya and Taj.

These hotels, being amidst the green forests, could not be seen distinctly earlier but now by chopping of all bushes from the surrounding areas, they have become quite prominent.

Further deep towards the East side, Nehru Park has been developed by the NDMC which has come up well. Hotel Ashoka close to it, had at one time been quite controversial on the ground that it had spoiled the green character of the area. Even now it looks predominant and can be seen even from Moti Bagh and Dhaula Kuan side.

The ridge since long has been a site, where heinous crimes have occurred. The planners even then proposed sites of Buddha Gardens and Mahabir Jayanti Park over there and for their development a total of 100 acres of the land was allotted; 80 acres for Buddha Gardens and 20 acres for Mahabir Jayanti Park. Is there any point in developing 100 acres of land, at a place so far away from the city and where even during day time, one would hesitate to go?

Is it worthwhile to spend such a huge amount for a few visitors? Along the ridge, close to the city areas, plenty of land is available which could have been utilised and named as Buddha Gardens or whatever. There was no need to destroy rocks and forests of the area just for entertainment's sake. After all a city has to be a functional one.

Was the placing of Rabindra Rangshala—an open air theatre over the ridge, thoughtfully planned; where shows are usually held during night hours? Can even a well-to-do person visit such deserted areas with family and that too at night, when hardly any bus or any other mode of conveyance is available. Though, marvellously designed by a Delhi Architect, this building has hardly been utilised properly on account of its poor location. It is being mostly used now by C.R.P.F., Army officials and their families who live close by.

Perhaps the Upper Ridge Road was basically carved out to link Buddha Garden with the city areas from the south as well as from the north side. Along it at Dhaula Kuan has come up a multi-storeyed building of military school and in future, more such projects are expected, which could be harmful to the ecology of the area.

The ridge portion in Karol Bagh area has unnecessarily been spoiled as there was hardly any need of roads—Rangshala Marg, Shanker Road, Link Road, Ganga Ram Hospital Marg, as other roads like Pusa Road, Jhandewalan Road, Panchkuian Road, Mandir Lane were already in existence to connect Karol Bagh with D.I.Z. area, Connaught Place and Pahar Ganj.

The ridge, which is at a higher elevation of about 20 metres towards the east of Ganga Ram Hospital, has now been badly damaged, as all over it multi-storeyed buildings have come up like that of—Sir Ganga Ram Hospital, Janki Devi College. Bal Bharti School, Gurdwara Nanaksar and others. Along the Link Road side also several petrol pumps and tall buildings like Jagjivan Vidya Bhawan, Tribal Museum, Springdales School etc. have come up which have completely blocked the view of the rocks in the area.

The Lakshmi Narain temple on Mandir Marg is a commendable project of an unique style as it is planned in harmony with the surrounding rocks and the greenery. Kali Bari temple is also an interesting feature of the area.

A narrow street from temple, called Lover's Lane, goes to the interiors of the ridge. The ridge in this area is the most beautiful and is 20 metres high all round, with deep valleys, where at place some yoga classes are taught.

One could still find along the lane horse tracks where horses galloped in the early British period, which go deep towards north and continue right up to Jhandewalan. Similar tracks are found on the ridge in the South direction at the back of the petrol pumps on Sardar Patel Road, which generally was used by the President's Body Guards and the early Britishers. It is said that in the British time the ridge particularly, was meant for horse tracking and that was how the site of New Delhi was discovered on Raisina hills.

In the West, the ridge extends to Todapur and Das-Garha villages (near Inderpuri), where its stretches are very low and beautiful. At Naraina also it is partly visible.

Similarly the north ridge is equally rich in terms of greenery and high rock sediments. At places like civil lines, posh bungalows are gradually converting into institutional buildings like schools, public authorities, hospitals, which as such could be hazardous to the ridge in the near future.

Rajpur Road and other roads leading towards Maurice Nagar and Jawahar Nagar are speedily becoming major thoroughfares.

The clearance of old Subzi Mandi was perhaps the beginning of the improvement of the ridge by which rocks and greenery of the area could be made visible and for this the credit goes to the late Prime Minister Smt. Indira Gandhi under whose direction this operation could be made successful.

Actually, costly development is not the only solution for saving the ridge. What we require is a perfect conservation and protection in the natural context and not in terms of the development of the ridge by laying lawns and gardens which otherwise look artificial and odd.

For saving the ridge we must concentrate on the many punctures holed in it, for which it could be said, that even single one among them is not desirable.

Talkatora Gardens could be made more beautiful, if its high boundary walls are removed from the west side, to open the view of the ridge forests. The single-storeyed buildings situated opposite Lohia Hospital that accommodate barracks and clubs need to be demolished as they block the view of the gardens. Around the stadium and the Regency Restaurant, a dense and scape may be planned.

The Willingdon Crescent Road could be built underground so that the greenery of the forest area in the west side could be blended with that of the President's Estate lawns and Mughal Gardens. Restoration of greenery around the Taj and Maurya hotels at Dhaula Kuan is also required to be thought off. The horse tracking on the ridge would attract a number of tourists.

The Simon Boliver Road could be closed and access to Mahavir Jayanti Park may be provided from Sardar Patel Road.

Efforts are required to be made to serve greenery of the ridge, from Link Road side as has been done at the site of old Subzi Mandi in the past. The semi-circular tank near Bhuli Bhatiari could be made open to public and activities like swimming, boating, fishing etc. could be started there.

Naraina

Naraina is the only place in Delhi where one is made to see and remember that the Capital is situated on the Aravali Hills. In the rest of the city, the sky-scrappers and high-rise buildings have effectively hidden this fact of Delhi's topography. The necessities of daily life have erased the natural environment for ever.

Complete restoration of the Delhi ridge is now impossible. But at least at Naraina and other areas where it is still visible something could be done if the authorities and the public so desire to save what remains of the ridge.

The first Master Plan of Delhi had proposed a ring road around Delhi without keeping in mind that this would destroy the ridge. Otherwise, the ridge would have to be cut throughout several places, such as Dhaula Kuan, Moti Bagh, Khyber Pass, Civil Lines and Mall Road.

In early times the Master Plans were crude piece-meal jobs with hardly a thought about whom it would benefit and what it would destroy.

The ridge profile at Naraina is a very inviting sight. It is of red textured stone that rises gradually from west to east to approximately 15 metres height, stretching horizontally straight to north and south to a distance of about one furlong. It then suddenly takes a deep easternly turn from the south where it is eclipsed by double-storeyed residential quarters of the defence forces.

On the north of the ridge lies the Naraina Village where

most of the residences have been converted into shops—an activity which has found its way across on the Ring Road.

The Naraina ridge requires immediate steps for conservation, preservation and beautificiation. Being close to the Palam Airport, it has a tremendous scope of developing into an important tourist centre.

It would not be easy to decide what kind of scheme would work. Delhi's climate, dry and semi-dry for most part of the year and also the chronic problems of water shortage, lack of open spaces and lack of water sports etc. makes it highly desirable to construct a lake at Naraina.

The lake could stretch from Naraina to Dhaula Kuan. The Ring Road would have to be altered into a loop towards west Delhi. It would be a tedious task since defence establishments have come up there.

The project could be executed in three stages. In the first stage, developments may be made exclusively in the area in front of the ridge facing the Ring Road. In the second stage, the double-storeyed residential blocks on south side up to the railway line may be demolished in order to open the view of the ridge. The lake could extend to this side as well. In the third and final stage, the scheme may extend up to Dhaula Kuan and beyond to the south, and a fly-over built at Dhaula Kuan. (The fly-over is under construction now).

The lake and its surroundings should be a giant landscape with feery services from Naraina to Dhaula Kuan. Facilities for fishing, swimming and boating may also be introduced. All around, the mini-forest could be developed.

The Naraina Village would be shaped into model village where old chaupals, old trees, wells, houses may be restored.

Kala Pahar

The Kala Pahar in Delhi, better known as Anand

Parbat—a range of Aravali mountains that extended from Zakhira Chowk in the West to Shadipur Depot in the South.

Certainly it must have been, in the past, in continuation to the Delhi's two ridges—North and South, as is evident from the bare existence of the rock sediments in the areas of Patel Nagar and the Sarai Rohilla. Though, there is no clear cut approach to the Pahar, but still one could reach over there from Zakhira side.

Hundreds of trucks, unauthorised colonies, railway tracks, clutter this area.

Quarrying operations over the years has destroyed the very beauty of this ridge. Many new industries and a cluster of residential buildings have added to the destruction.

The residents, it is believed, have settled there about 20 years back. And administration had never bothered to check this infilteration. The streets in this area remain clogged with men and machines. Hardly four feet in width they lie as much as 15 metres below the road level.

The complete restoration of the Kala Pahar can be ruled out. An attempt could be made to blend it out of the Delhi's south ridge which could be possible from Patel Nagar and the Pusa Gate side by suitable planning. A massive programme of afforestation need to be undertaken to achieve a green sky line, as this would serve dual purpose of absorbing smoke from tall chimneys and secondly would beautify the area.

The industries should be told to take pollution control steps to restrict the growing pollution of air and water in the area.

The fodder Mandi at Zakhira should be shifted to another alternative site alongwith the idle truck parking and thus the ridge could be made more open. The residents of the unauthorised colonies too should be moved to alternative accommodation so that the ridge profile could be made more visible.

To further develop this neglected ridge it would be most suitable if beautiful meadows could be developed along the ridge from Zakhira to Shadipur Depot and at Anand Parbat side where sheep, goats, deer, horses and other animal varieties could be introduced. Besides, these parks and open spaces can be developed for recreation of the local residents. A mini-forest could also be planted to act as a reservoir of oxygen. The depressions in the rocks could be turned to lakes and ponds.

A Master Plan of the Kala Pahar is desired to be prepared in regional perspective by the experts in this field and every effort should be made for conservation of the rocks, the trees, the birds etc. of the area.

Search of Peak on Aravalis

The Bounta hills near Delhi University, are beautiful ranges of Aravalis which remain almost lost and forgotten, as allround them, have come up busy roads, and multi-storeyed buildings like that at Malka Ganj, Kabir Basti, Civil Lines, Delhi University, Timarpur etc. which have completely blocked up its view. However, it is only from the side of St. Stephens College that one could have a glimpse of it. The ridge in the past is believed to have touched Yamuna, in the north easterly direction as is evident by Ashoka's Lat which lies near Ajitgarh and for which it is said, was brought by Firuz Shah Tughlak to this place from Meerut through Yamuna river.

The ridge has been divided at several places by the surrounding road network which connected various buildings built over these rocks. At Malka Ganj on the west side, a number of residences have come up just on the face of it which thereby have completely blocked it's view. The slum area called Kabir Basti, where mostly potters live, has sprawled all over. The adjoining roads, create severe traffic chaos which disturbs the calm and quiet environment of the adjoining green spaces near Ajitgarh and old Subzi Mandi. Hindu Rao Hospital—one of the oldest in the capital, where only dog bite cases were earlier attended, encroaches a large acreage of land over the ridge and has thereby spoiled one of the highest peak of the Aravalis which probably like Shillong peak, could have been named as 'Delhi's peak' and it could be an important tourist spot—being one of the highest elevation on Aravalis, from where almost entire Delhi could be seen. Moreover, Ajitgarh and Ashoka's Lat could have been the additional tourists attraction of the area.

The South side of the ridge is getting out of sight as several transporters offices have come up there. Further, the Bungalows belonging to MCD and Railway officials, opposite Tis Hazari, block the ridge view and the greenery of the area. In the Civil Lines area at Court Lane, Rajpur Road, Flag Staff Road and at under Hill Road; nursing homes, schools, offices and other such institutions are now functioning which invite heavy crowds and intensive building activities.

The ridge in continuation in North-easterly direction was broken by Mall Road and Alipur Road and the detached portion lies still in Timarpur area where continuous blasting operations are in operation. All trees have been cut there and the area has become a large dumping yard.

The ridge has now been re-named as Kamala Nehru ridge and developments have already taken place at Bounta and at Choburja Park. But did this at all help the common man? Of else is it helpful in restoration of the ridge in terms of its rocks and greenery? Certainly not. The development has infact spoiled the natural tint of the area. The Choburja Marg that leads to Tis Hazari and to ISBT, attracts a large number of vehicles, which create air and noise pollution. The

Bounta and some other rest houses, on the ridge which are the left out traces of the old mutiny days, look excellent. The railway track over the ridge which earlier is said to connect Old Secretariat with Mori Gate Railway Station, is however, not traceable.

The strategy to develop the north ridge would be to make the ridge view more open and to remove noise and air pollution from the area. At Malka Ganj, adequate building surgery is required so as to open the ridge view. The residents of Kabir Basti may be shifted and alternative accommodation be given to them.

The construction of multi-storeyed buildings in the area be banned. The Government offices like the Authority offices should be shifted to some alternative place. The schools and nursing homes could, however, be permitted.

As per Mohan's concept near Hindu Rao Hospital a viewing gallery and a tourist resort could be planned, as that would be Delhi's peak. As such, this would require more exploitation.

The network of the roads surrounding the ridge should be re-examined. Alipur Road could be built underground so as to blend the ridge and the river. A master plan of the ridge may be prepared and efforts may be made to link north ridge with other isolated parts. Also extension of north ridge up to Yamuna in north and to New Delhi ridge in South be examined. All monuments and ancient buildings in the area should be preserved.

10

River Banks : Yamuna River

All over the world, and throughout history, human settlements arose mostly on the banks of rivers. The world's first civilisations developed along the great rivers: the Tigris-Euphrates in Iraq, the Hawang Hou in China, the Nile in Egypt and the Indus in what is today Pakistan.

The river Yamuna has long since been associated with Delhi. This is clear from the famous epic, *Mahabharata*, which mentions a city called Indraprastha built along the river Yamuna. Yet why are we shunning this great river today?

In history there is a reference to the city spread between the Nigam Bodh Ghat and the Purana Qila. Subsequently on this western bank rose capitals of other great dynasties and empires. The Yamuna today flows farther to the east, as is evident from the fact that the graves of Saint Turkman and Razia Sultana, who were buried on the river bank, as mentioned in history, are however, now found at the present Turkman Gate site. Apart from its navigational potentialities and as a defence barrier providing protection for those who ruled Delhi, the Yamuna was a precious source of recreation, entertainment and natural beauty. This fact is amply borne out by the existence along the river banks of Tughlakabad Fort, Ferozeshah Kotla ground, Purana Qila, Humayun's Tomb and the Red Fort.

A 1914 map of New Delhi, clearly shows, how Sir Edwin Lutyens had planned to locate the new imperial city along the river by proposing a big reservoir around India Gate and ultimately joining it to the right front. Unfortunately, this concept was never realised.

In the post-Independence period, Delhi grew haphazardly. A large number of labourers and industrial workers found shelter in the trans-Yamuna areas. The Gandhi Nagar, Krishna Nagar and other similar low-income neighbourhoods sprang up along the river bank. The old railway bridge was followed by other bridges of Wazirabad, Nizammuddin and Indraprastha. On the Western bank rose colonies like Maharani Bagh and Friends Colony. When Ring Road replaced the Old Bela Road, the river front started coming closer to Delhi.

In the eighties, however, the trans-Yamuna areas began to attract the upper classes, when a number of plots for more highly placed persons were developed. The NOIDA complex too, started drawing small sections of Delhi's population. The Indraprastha estate became a major link between East and West Delhi.

The river front in Delhi has remained a relatively neglected part of the urban area. Its recreational potentialities have yet to be realised and exploited. We need to create pleasant walks along the entire river front from Wazirabad and beyond, up to the Okhla weirs. The proposal to link the deep city interiors to the river front need to be carefully examined.

The Master Plan for Delhi proposed the development of both the banks of the entire river front of the Yamuna. It could be an important recreational area for the city. Among the proposals are national playgrounds near south of Rajghat, bathing ghats, swimming areas, tree-linked boulevards and parks.

The development of the river front is partly dependent upon the physiography of the river: its location, course, type and its volume of water. It also depends upon the economic feasibility and scope for utilisation of land on either side of the river. No worthwhile effort has so far been made to plan and develop the Yamuna river front. Suggestions are now being examined to develop the entire river front scientifically.

For study of the river front the entire stretch of Yamuna at Delhi has been divided into four parts.

This first part deals with the river front area between Wazirabad to ISBT, second from ISBT to Pragati Maidan, third from Pragati Maidan to Humayun's Tomb and the last from Humayun's Tomb to Okhla.

Wazirabad to ISBT

The Yamuna front from ISBT to Wazirabad, which happened to be quite natural and touched Delhi's north ridge in the past; lies in a state of a big mess now on account of the construction of multi-storeyed building of ISBT which not only damaged the scenic grandeour of the river bed and eclipsed the beautiful city walls at Kashmere Gate but also spoiled the greenery of Qudsia Gardens. The over-bridge at ISBT which is under construction and which no doubt, would link east with west, but at the same time, would be a big hindrance in the development of the river front and would simultaneously be creating air, noise and water pollution.

The civil lines zone, which is a large encroachment over the river bed, looks like a long dead wall raised against the river front. The tall building of Budha Vihar Monastery, that came up in early seventy's, closed not only the river view but also has checked the expansion of the river bed. The other encroachments like that of cowsheds, wrestling dens on that side look very ugly. At Metcalfe House, too, multi-storeyed buildings have come up, which do not appear to be healthy proposals.

Along Mahatma Gandhi Road, two large Jhuggi-Jhopri Colonies, namely Aruna Nagar and Old Chandrawal have come up, which occupy a large area. Adjoining the Gurdwara shrine at Majnu-ka-Tilla, ugly dumping waste sites are coming up. The adjoining building of the Shrine of Majnu-ka-Tilla now remains surrounded with noisy roads.

The Najafgarh drain is a big nuisance. The sanitary land fill sites near Timarpur attract vultures and other such birds, not good for the city. At Wazirabad barrage, a large area has been swallowed up by the MCD offices of water works department and by other unauthorised dwellings. However, near Wazirabad and Jagatpur villages, Yamuna front is extremely beautiful and attractive. Further deep i.e. near villages—Burari, Nathupur, Ibrahimpur, Alipur and Palla the river front is again spoiled as large number of illegal constructions have started coming up.

In the trans-Yamuna side, the river front position is too bad, as all over unauthorised constructions have come up all along the marginal bandh at—Shastri Nagar, Usmanpur, Daljit Nagar, Kartarpur, Ganwari, Garhi Mehndu, Bhajanpura, Khajuri, Tukhmirpur, Dayalpur, Karawal Nagar and at Subepur which as such occupy a major city area. Even constructions are coming up below the bandh in the flood belt zone.

A concrete proposal is required for the development of the river front in terms of conservation and protection of the river bed in view of the Yamuna's past position. One of the proposal for 21st century could be to start navigation from Palla to Okhla, as this could help in making journeys easy and comfortable. To solve city's drinking water and power needs, more lakes and canals could be developed along the river, for which vacant sites at Jagatpur, Palla, Subepur, Burari and Usmanpur could be examined.

For making the river front more livable for tourists,

temporary tourist resorts could be built, all along the river, at subsequent intervals of 5 km. At Majnu-ka-Tilla and at such other places where river fronts are extremely beautiful and which are in easy access, major recreational centres could however, he developed.

The pollution of the river may be minimised. The wastes of Najafgarh drain could be treated with highly scientific technology by which even the waste water could be recycled for drinking purposes. The sanitary land fill sites at Timarpur require urgent shifting to some other alternative places. For checking the contamination of water, proper chemical laboratories could be set up at Palla and at Okhla.

The bathing ghat sites could be shifted from ISBT and developed at Wazirabad and at other places along the river where water is not contaminated. The efforts are required to be made to arouse proper sanctity among the public for the river, so that the water could remain more pure. Few temple sites could also be proposed over there.

The traffic conditions in the area could be solved by closing Ring Road from ISBT onwards and Mahatma Gandhi Road's traffic could be diverted along Alipur road. The latter, however, could be widened and shifted slightly towards east side. The river bed may be extended up to Kashmere Gate, Mori Gate and up to the ridge, as was in the past. All building activities be totally banned in the river front area.

The vast open flat land, lying opposite to Civil Lines along the river, could be utilised for indigenous sports activities. In the trans-Yamuna side also, such activities may be repeated. This would attract lot of people who would go there to watch games like kite flying, wrestling etc. and this as such could become their main entertainment source.

From Khajuri to Subepur in the trans-Yamuna side, the proposal to start domestic animal farming and fish farming be examined for which suitable *mandis* may be set up in nearby areas, for sale of products like milk,wool, meat, fish, oils etc.

The alarming signal of the growing unauthorised colonies, coming up below the marginal bandh level at Shastri Nagar and at Khajuri require special attention; for removal to other alternative site. The beautiful lake developed at Usmanpur that even attracts migratory birds could be conserved and protected and made open for tourists and for the public.

Around the over bridge at ISBT which is under construction, extra care is required in terms of the river bed conservation.The proposals to start boat clubs, flying clubs, race course along river front be examined to make such aristocratic games popularised among the general public. For an ideal river front, it would be better if an international design competition could be floated among artists and architects.

ISBT to Pragati Maidan

The Yamuna front at Delhi, from Pragati Maidan to Inter-State Bus Terminus which happened to be a very lively area of the past, as there stood along it, beautiful forts and temples, is now the most congested area of the city as all over multi-storeyed buildings, power stations and unauthorised hutments have come up.

The Indraprastha thermal power station is the greatest pollution hazard of the area. Surprisingly, it is again being expanded towards the south east. The site of hot mix plant situated near the power house creates pollution of air and water.

The power house at Rajghat is another pollution hazard of the area which generates little and expensive power. The MCD and DESU offices located that side are big eyesores. The mushroom growth of hutments near marginal bandh at Kanchanpuri spoil the river bed.

For the Samadhi complex—Rajghat, Shantivan, Shakti Sthal and Vijay Ghat, no efforts have so far been made to bring the river scape close to them, even though, the river flows closeby. Moreover, extensive greenery at Shantivan is over dominating and does not match with the river landscape. The Pantoon bridge that side, along which a nallah drains to Yamuna; disrupts the continuation of the river front.

The Salimgarh fort which at one time was accessible by boats only, remains now deserted as crowded roads and over bridges have come up all round it.

The old steel bridge has almost outlived its utility. Over the bathing ghats, located near by it, the water remains contaminated and is not fit for bathing purposes. The Nigam Bodh Ghat creates a passive feeling in the environment. The marginal bandh over there blocks the river view. The overbridge which is under construction at ISBT would be spoiling the beauty of the adjoining Qudisa Gardens and the other green spaces.

At Delhi, Yamuna has its best fronts in the trans-Yamuna area at Kailash Nagar, Gandhi Nagar and Krishna Nagar, where the river flows quite close to the city. One could get from there a beautiful view of the silhouette of Red Fort and Jama Masjid and sight of the crossing trains over the over-bridges. Such spots do exist on the west bank also, near the boat bridge and near the over-bridge on southeast side of Indraprastha Power House. The recreational centres could be developed around such spaces and a toe path connecting these centres could run all along the river front.

The recreational centres may have the facilities of water sports activities where international tournaments could be arranged. This, as such, could attract the tourists. Besides, there could be activities like fishing, boating, swimming, etc.

The proposal to install lakes at low lying places and possibility of installation of mini underground pollution less power plants be examined to solve the drinking water and the power needs of the residents. One such spot, where the lakes could be constructed lies at Krishna Nagar between the Brijlal Goswami Marg and Raja Ram Kohli Marg.

Extensive plantations are the least desirable along the river front. Rather a river bed would look much better in its natural form, colour, shape and texture. The sandy beds amidst water lagoons and melon crops form an excellent combination. The floods could only help in achieving natural, wider and thick sandy beds.

The riverscape may be allowed to penetrate the city from the three points; from Delhi Gate to Turkman Gate along Ramlila grounds; from ISBT to North Ridge along Qudsia Gardens and Mori Gate and finally from Shantivan to North Ridge along Red Fort, Jama Masjid and Queens Gardens.

The office complex situated at the back of Rajghat including the power-house should be shifted to some alternative site to have a better view of the river front.

The marginal bandhs could be landscaped by possibly developing water bodies alongwith them. The plying of heavy vehicles over them be totally banned. When it is to be used as road then the road level may be raised sufficiently high to have a view of the surroundings.

The existing bandhs which can sustain 2.5 lakh cusecs of water may be redesigned for 5 lakh cusecs. Proper meandering of the entire river bed is required as anti-flood measures.

The air pollution created by the Indraprastha thermal power station may be minimised by applying electrostatic precipitators along the chimneys. The undesired expansion of power house particularly towards southwest side be checked to protect the river front.

The expansion of the river bed up to Mathura Road, covering thereby the entire Pragati Maidan be examined to provide relief spaces to the ancient monuments of Purana Qila.

For bringing riverscape close to Shantivan, a third additional bandh could be build close to the river along which a lake may be developed which may touch Shantivan and Shakti Sthal.

The approach to Samadhies could be provided from the river side, rather than from Ring Road as in that case one would get an additional beautiful picture of Delhi city that is, a silhouette of the monuments; a close view of the river and finally the sight of the city panorama of Shahjahanabad and Shahdara.

The existing bathing ghats near Steel bridge may be shifted to an alternative site, where water is less contaminated. Nigam Bodh Ghat may be extended towards north and across Yamuna to meet the requirements of the city. The expansion of the Yamuna front on that side up to the city wall may be examined.

The mushroom growth of huts at Boat bridge; at Kanchanpuri near Rajghat and at Shastri Park extension across GT Road may be removed and the residents be given alternative accommodation.

Opposite Shakarpur and at Laxmi Nagar where the river bed is too wide and extensive cultivation is being done presently; proposals to start water sports activities opposite the stadium complex and fishing development projects opposite the power house may be examined.

The water line as was in the past may be restored and be made to touch Red Fort, the Feroz Shah Kotla ruins, Salimgarh Fort and Darya Ganj by developing suitable landscape surgery or by creating water bodies.

Pragati Maidan to Humayun's Tomb

The river front area is considered somewhat as a waste land, where all undesired activities, otherwise hazardous to the city areas, are usually proposed.

In Delhi, particularly in the area lying between Humayun's tomb and Pragati Maidan, a mess has been created. Sanitary land fills have come up in the area which emit foul smell, look filthy and attract vultures and birds harmful for the city residents.

Moreover, hot mix plant sites are in regular operation which create air pollution. The boiler wastes from the adjoining Indraprastha power-house is being dumped over there which has made the entire river bed look ugly.

Besides the Hazrat Nizamuddin railway station, located on the face of Humayun's tomb, the railway lines laid over high embankments combined with the railway overbridge at Pragati Maidan, have altogether spoiled the beautiful sky line of the chain of the centuries old monuments of Arab-ki-Sarai, Humayun's Tomb and Purana Qila, which otherwise in the past, enjoyed a perfect setting along the river bed.

The multi-storeyed gurdwara building which has just come up, has blocked the view of Humayun's tomb. Originally it used to be a single-storeyed structure.

Due to the lack of control over the river front, a large number of unauthorised extensions of Nangli Raza and Kale Sarai villages have come up at Nizamuddin thereby blocking the view of Humayun's tomb. At Kale-Sarai, a pond has been developed which has come up well. Towards north of Kale Sarai lies a nallah that drains into the river and to further north of it are situated housing clusters of Siddhartha Enclave which again block the view of the river and of Humayun's tomb.

The monuments and the archaeological sites at Purana Qila, some of which even date back from Pandava's time, as is revealed by Bhairon's temple located over there, have totally been got eclipsed by the network of the roads and the railway lines laid at Pragati Maidan.

The Ring Road particularly from Pragati Maidan to Humayun's tomb, appears to be superfluous, as it serves only to connect the west of the city with the east through an over-bridge across Yamuna. The euclyptus trees there not only give an ugly look but block the river view too. A few rag shops which have come up there do not look healthy.

The eastern river front is comparatively much wider, than that of the west. Along the marginal bandh which runs north and south, a good road has been built towards south which leads to the areas—like Patparganj, Mayur Vihar, Samaspur and Noida whereas on the north no such road could be built, as there is an unauthorised colony called Pandav Nagar that extends up to the railway lines near Shakarpur.

The extensive cultivation along the river bed, which was done earlier by the farmers of the adjoining areas is now suffering, as from there the earth is being removed by prospective building contractors, with the result, giving the river bed an ugly look.

For developing the river front first of all the restoration of the water line, along Humayun's tomb, as was in the past, may be thought of. The perfect conservation of the monuments, greenery and of the open spaces is earnestly required. All kind of building activities in the area must be totally banned, otherwise this would create air and water pollution.

The sanitary land fill sites and the hot mix plant site should be shifted to an alternative place. The dumping of boiler wastes from Indraprastha power-house, could be switched, preferably towards the east side of the power-house itself, along Yamuna or it may be converted to bricks for construction purposes.

The closure of the rough passage on the Ring Road may be examined and if possible, the link to the Yamuna overbridge at Nizamuddin could be made by an underground road, which may start from the Mathura Road and Zakir Hussain Marg crossing. The euclyptus trees growing along the road should be removed.

The network of the railway lines which run at the back of the monuments towards east could be made underground so that the view of the monuments gets opened from the river side as was in the past. Hazrat Nizamuddin railway station could be made to serve as a local station and if possible this too may be built underground.

The unauthorised extensions of villages may be regularised and no more constructions be allowed there.

The nallah draining into the river be cleaned first at least 2 km away from the river and the gurdwara building standing at the back of Humayun's tomb, may be provided an alternative site.

The road network along Purana Qila and Pragati Maidan be re-examined for closure, so that more relief spaces could be provided to the monuments over there. Instead, a parallel link could be provided from the Indraprastha bus depot side.

At the back of zoological garden a few recreational spots like bathing ghats could be developed.

Along the Humayun's tomb, a 60 m wide and 40 m deep canal could be dug which may fulfil the drinking water needs as well as the recreational requirements of the residents. Towards its east, a bird's sanctuary may be developed which may run north to south.

Identically on the eastern river front, a similar canal may be developed with a fishing pond alongthe recreational spots and special zoological gardens situated on its west. Adequate afforestation measures may also be taken up there. A high railway embankment towards north could be built either on stilts or underground so that the river bed at Shakarpur too could be got blended with these areas.

The intensive cultivation operatioons in the area could be switched on to the production of aqua-vegetation for which Delhi is too much deficient. The development works should look natural as that would be more impressive.

The fishing ponds and pets rearing farms, would not only look interesting but also would bring high remuneration to the city. The bird sanctuary is bound to attract more tourists.

With the completion of the schemes drinking water needs of the Capital would be sorted out and also power problems would be solved, if somehow, the scheme of laying mini underground power plants becomes successful.

With all these proposals, it could be hoped that the capital's weather conditions would improve coonsiderably and an ecological balance would be achieved all over.

The possibility of introducing river scape inside the city up to the north ridge from areas like Humayun's tomb, Delhi Golf Course, Lodi Gardens, Race Course, Nehru Park may be examined, so that these open spaces too get blended among themselves.

With the dug up earth from canals construction, big mounds even as tall as mountains, with varying sky lines could be built all over the river bed and which with proper setting and orientation could definitely help in checking Delhi's wind and dusty weather conditions. The proposal of tarnsplating the overall environment of lakes like Bhim Tal and Tali Tal of Nainital could be examined by suitable landscape surgery which is of course possible.

Humayun's Tomb to Okhla

The Delhi's river front that stretches from Humayun's Tomb to Okhla could though, have been very popular and attractive because of its being very natural, neat and clean yet, it is found to be in the most unsatisfactory and neglected state, as allround it, have come up residential areas, some of which are, however, posh like New Friends Colony, Kalindikunj, Maharani Bagh whereas many unauthorised colonies too have come up there at Nur Nagar, Taimur Nagar and Khizrabad at Okhla.

It is, however, only near Okhla village that a picnic spot could be found where, though canteens etc. have been provided but the overall development is not at all satisfactory as there is no other activity like boating, swimming etc. for the common public. It is only for the Army people that a few boats have been deployed there in a near by Club. The Agra Canal too is hardly impressive at its source.

Near to theKalindi Kunj an overbridge which has been built up recently to connect NOIDA and other trans-Yamuna areas, the river front is quiet wide, natural and is full with thick sandy beds. A lake, however, developed that side does not look so healthy a proposal.

The large encroachments of Siddharth Extension which have come up opposite to Maharani Bagh are spoiling the river bed, as allround, only tall buildings could be seen. Similarly near Jamia Milia University and at CRRI Complex identical experiences are met with. Near Sarai Juliana and Sukhdev Vihar we again came across the residential buildings that block the river view. The nallah running at the back of Kalindi Colony is a big nuisance for the area.

The development of the river front would require firstly

cleaning of the unauthorised colonies which are fast coming up near Okhla and at other places. Right from Humayun's Tomb to Okhla, for both banks, a perfect development plan is required to be prepared which should have sufficient pedestrain access from the respective city areas. The Development Plan in the initial stages, may only involve conservation of sandy beds, water lagoons along which pedestrain paths could be provided. The facilities like that of swimming, boating etc., however could be provided all along the river to make it attractive. The horse riding and such sports activities could also be introduced over there.

Any more intrusion of tall buildings in the area should be completely banned. The pollution of the river should be minimised. The chemical laboratories could be set up at Palla and Okhla, to test the degree of contamination of water, as is important as per the statement made by Prof. Arceivala that Agra may be drinking Delhi's sewage which is a well documented analysis of the river along 195 km down stream to Agra, revealing thereby an increasing deterioration in the quality of water. Also tall viewing towers could be put up near Okhla which could act as a watch tower for the entire zone.

From Okhla to Mathura, recreational Centres may be developed along the river, at subsequent intervals of 40 km or so where facilities like restaurants, medical aid, tourist enquiry etc. could be provided.

11

Wild Life

The Delhi citizens might remember that in the past, how often the newspapers covered the items like Hayena creating terrors inside the city and that too in the area where VIPs lived. Probably the animals came from the nearby ridge forests. This illustration shows how rich was the wildlife of Delhi which now unfortunately has almost vanished. The rich fauna constituted several varieties of animals and birds like—fox, jackals, hayenas, wild pigs, leopards, Indian gazelle, hare, more rats, bats, wild dove, barhminy kite, pigeons, cuckoo, bulbuls, tailor birds, larks, strokes, oriols etc.

The wildlife at least in Delhi has suffered a great blow on account of various reasons. A number of animals specially frogs are killed daily by the pre-medical students just for the experiment's sake when the ratios of those who really enter medical profession is hardly 2 per cent. The environmentalist have objected to it and have desired to set certain restrictions over it. Again the animals like rabbits and others are being slaughtered and many are left in a pitiable condition just for the preparation of the cosmetics and drugs. Moreover, animals are being butchered for meat and eggs are being used for food purposes which as such are retarding the cycle of the animal lives.

The rapid urbanisation is responsible for the fading

away of the wildlife in Delhi. All over the ridge blasting operations were carried out by which most of the animals and birds ran away. The construction of roads and buildings too have adversely affected them.

The cutting away of trees and destroying of orchards specially from western Yamuna canal and the ridge areas affected particularly the bird lives as they did not find any proper place to make nests or any fruit tree over which they could sit and enjoy.

The drying of the water bodies mostly ponds and lakes too, are the factors responsible for affecting the wild life as those were the places where the animals and birds used to relax during hot hours of the day. The migratory birds that come in the winter season every year are also being affected. We could, however, find them in large number at Yamuna or at lake like at Usmanpur village in the trans-Yamuna area and at other places during the season.

Often near Delhi ridge we might have come across bird watchers, stopping their cars and rushing to the nearby trees just for curiosity sake to check whether that is the same bird whose sound they had just heard. Mr. Khushwant Singh a noted journalist and an environmentalist too gave very interesting talk on the television about the bird species of Delhi ridge.

The zoological garden at Delhi which in early seventy's had no more than 20 birds and animals is now having a big collection. Of course special sections like snakes rooms etc. have not yet been started there.

In individual homes pets like dogs, goats, sheep, horses could be seen. Still vendors could be found selling leaves of mulberry, *piple* trees—a food of sheep and goats on the streets of the walled city. In posh areas veternary doctors are running good business.

Conservation of the wild life is very important for which it is ern'st while desirable to protect all natural elements like trees, orchards, ponds, lakes which are for their development. The animals farming projects may be taken up particularly for the dying species to quickly make up the deficiency. Every kind of help required, be provided in this direction. These could prove as economical assets also as animal skins and other body parts are now fetching high rates in the foreign markets.

With the forest, like that, developed at Hauz Khas, the wild life has been little bit protected. There the peacocks are usually found dancing on the terraces and roofs of the adjoining residential buildings. The lake which is yet to be developed may attract a number of birds. The deer kept in the deer park over there, look beautiful.

There could be more pet shows held inside the city to arouse the public interest towards protection of wild life. The bird shows as is common in China could be popularised in India too. The seminars, the film shows and the exhibitions could be arranged over such topics. The industry should be made to be economic-oriented for which efforts may be made to produce rare kind of animals and birds not accustomed with this climate for which adequate experiments in this direction could be taken up.

Mohan's concept suggests the industry's expansion in a large way. At micro level, even in small neighbourhood parks, say of the size of 200 m^2; animals like deer in two or three numbers and a few birds could be kept. In the bigger open areas even panthers could be kept. Thus, all kinds of animals and birds could be displayed in a colony.

For this of course management has to be efficient. Its advantages would be manifold. Firstly the wild life industry would expand. Secondly the children from the very beginning will have a touch of it and a regard for it. Thirdly these areas would become as self-developed picnic spots which are required also in the context, as city has been expanding and where it is difficult to go to the far off areas for recreation. Fortunately this would act like a big exhibition where often guests could be taken. And lastly, this could prove as an economic asset for the Government of India in earning handsome forign exchange.

Global Casualities

More than 440 tons of dead eels have beenfished out of the Rhine riveralongwith front,white fish and pike when Sandoz factory outside Basal, Switzerland caught fire. In Africa's Victoria Lake more than 250 species of fishes are threatened. The haplochromines a native fish has been reduced to almost nil say 10 per cent from 75 per cent by nile perches introduced as an misguided perception. Japan's methl mercury poisoning during 1950s and 1960s left hundreds of fishes dead and thousands crippled in Southern Japan.

12

Environment and Habitat

Seeing present day housing, urban or rural, don't we feel, that the quality of living, some how, is degrading, may be because of the economic constraints or due to lack of the technical knowhow or for some other reason. A time was, when in villages even, though, there might not be available two times a meal in a family but the home could never remain ignored and it was as neat and clean as possible; with cow-dung plastered and so on. Even separate residences for the winter and the summer were planned and for the living, the separate *Baithaks* (drawing room) were provided. Could we afford all this now, in the present day context?

In those days, the local building materials; mud, wood or stones were, of course, freely available and moreover were free of cost. Only the labour was required and for which generally the male members of the family volunteered with the help of the Scheduled Caste people of the village; called *domes,* and erected most of the structures, the walls, the floors, the roofs and built windows by cutting trees and sawing wood.

The residences were symbols of power, caste and dignity; and even from their looks, it could be easily recognised as to which building belongs to whom.

In busy cities, are often found urban agglomerations;

where people live in slums, ghettoes, inthe worst possible environment, not fit for the animals even. The illegal settlements get born which remain as permanent eye sore for the Municipalities. The extensions of urban villages again find the identical second rate treatment from the Authorities.

The Housing Boards and the Development Authorities though, are the only responsible Authorities to care for the housing; but unfortunately they are found to be exploiting the common man by creating deliberately a dearth for housing by building their monopoly and doing extensive publicity and ultimately not responding accordingly, for one reason or the other.

There are no researches on housing aspects and whatsoever is got done is only one sided efforts without any crit. The public does remain groaning afterwards; cursing the authorities for all odds, for which the Authorities have least regrets and boldly come up with the announcement, that one who is not satisfied with the house, could surrender that to them and can take back his money incurred over that. But would any prospective buyer appreciate such bargain, when he could easily get quite a high premium, for his flat or plot in the open market. So such confrontations remain in the housing market.

Before us we have the examplary trends set up by our great Master Architects; like Frank Lloyd Wright's falling water; Le-Carbusier villa Savoy Poisey and many such others. The Indian *Haveli* culture too is not less impressive. It is very gaiety and is full with life. The open internal courtyard invites all social and family activities and serves multi-purpose functions. Unfortunately such concepts now remain totally ignored in the present day housing concept.

As land is very expensive, therefore, its utilisation is ought to be maximum. But it does not mean that it should be at the cost of the open spaces or natural left out areas like the river and the ridges. The land situation in urban areas is very quarrelsome which often invites legal suits among the individuals or with the Authorities. For acquiring land, the Attorneys, the Benami transactions and other underground methods start getting adopted specially for evading taxes. The relaxation of urban land and ceiling act could check city expansion. Mr. Romesh Bhandari, Lt. Governor, Delhi is believed to be taking interest in the subject.

The practice of developing sectors planning, leads to monotonous housing that present an artificial environment, not like the organic development. If there is a road, then it would continue 4 to 5 km without any disruption and there is no awareness of the urban serial vision that such long vistas look vey odd and as such require to be broken up after every 1.5 or 2 km by proposing the round abouts, or such features over the central axis. Do we not observe such concepts in Lytten's Delhi how India Gate disrupts the movement of the long straight road. How the Fatehpuri mosque comes in way of long stretched Chandni Chowk and how in Tokyo a new Tokyo Plan 2025 has been conceived by Kisho Kurokawa on identical concept. Of course Chandigarh and the Gandhi Nagar are examples of sector planning. We do come across some deficiencies despite there being plus points. The Planning of DIZ area in Delhi was very healthy as it was on a small area and the sectors too were very small and easily managed. The roads still after 50 years are as spacious as were in the past.

The facilities which are part of the housing complex are seldom found to be to the utmost satisfaction of the residents. Whether they are roads, parks, shops, schools or are of basic infrastructure nature like-sewerage, water, electricity, etc. all require priority consideration for the success of any housing scheme. These residential cities are to be designed not for only 5,10 or 20 years to come but should cater for 100 or more years. Is it not displeasing that the public still for better facilities is going to far-flung central places for getting better quality and lower priced housing and services?

The different LIG, MIG, HIG groups, no doubt, sound well but don't we realise that each group gets the amenities accordingly, as per the income range when a human being is the same. Why such partiality is then there?

A craze to own a house gets born in an individual's brain from the day, one joins service, otherwise in the past the people often left for their home towns after retirement. It may be because of the high rentals or non-availability of houses or other such reasons. The individual's survival in a busy city is, of course, difficult because a major part of the salary would be sacrified for the house rent only.

More researches are required to be done for finding out a perfect solution for which the architects, the artists, the psychologoists, the sociologists, the economists all are required to sit together with an open mind and evolve a practical nationalised housing policy whichmay suit the common man.

The various Authorities responsible for housing must exert more to find early solution for housing. They may either start five year plans or Point Programmes could be advocated for this. Of course, efforts of Mr. S.K. Sharma, former Chairman HUDCO in this direction are worth commendable.

The Public participation is a must for making housing programmes successful. For this, mass media like TVs, Radios, Newspapers are required to be catchy and should come out with the stories concerning good living environment. The seminars could be held for it. For maintenance and legal guidance 'Public Bureau' could be started.

The Trans-Yamuna area has been the worst zone so far in Delhi. A case study of it describes its major agonies due to illegal settlements, slums, urban villages, etc. The breaking out of cholera in May-June 1988, had been the major discussion topic that covered the major headlines of almost all the daily newspapers. Though many new developments in the form of regular societies or Group housing have come up but the level of infrastructure is very poor. Hardly there is any good open area except little bit along the Yamuna river. The road network too is not satisfactory and there is cute traffic problem. The mass transportation system is worse. The Noida complex is inviting heavy influx of population. At this stage, if no new Master Plan is planned, this area will be doomed and would create the worst urban chaos in the total region.

As per Mohan's concept, after a man occupies a house there should be no liability on him for paying any fees to any authority like-house tax, lease money or even for maintenance or as association fee. All this be either arranged initially at the time of sale and purchase or should be discontinued for ever or be borne by the Authorities themselves.

13

Implementation of Total Quality Management

The practice of rolling of a firm's product which is accepted blindly in the society is an indicator of the quality of the product to a large extent wherein little credit goes to the publicity that must be there. The quality management has remained in adoption since the beginning of early times when big industrialists were themselves going to the retail markets to enquire about the sale and quality of their products and if at all any shortcomings they found they either ordered for withdrawal of the same from market or made such alternative arrangement to relieve the retailers. For Lala Sri Ram, this was mostly said that he adopted such an attitude while running DCM and Sri Ram Mills.

May be the management was not theoretically set in that line but when the head was involved and in case we analyse his activities from the survey; then all present management principles could not be matched with them as present management has not added any such superfluous item or miracle.

There is a tough competition in the market about each item whether smaller or bigger. Besides publicity; anti-publicity margin by competitors can also not be overlooked as that hampers the sale of a product. Moreover, the reailers are interested in taking those items where their margin of profit is comparatively more. The consumers are now not ignorant about a product and before one goes to the market, a survey is already done by him.

The various quality indicator agencies even governmental have come up to mark a product but they do so in the beginning only of a sample. Rest of the items that are sold in the in market totally goes unchecked until some sample is collected from them. And even if a sample is collected and if that is not up to the mark, that can be done so through underhand methods. The lust of entrepreneurs to earn more and more to boost business by giving more commission to agents is really the backbone of degrading the quality of the products. There is no nationalist spirit and if any consumer goes to the court, his case can easily be twisted.

The ISI and other nodal agencies for marking quality products can not rule out absolute quality or even standard but can assure tolerable one.

Now-a-days consumers are intelligent and they weigh the materials, read the labels well, see if prices are marked and given to them at reasonal rates. That is in tune with bulk purchasers but for petty purchasers hardly these points are taken care of and God is the only saviour for them. In case the product is not good there is no one to take that back.

In U.K. such practices of consumerism are very tough. The retailers are very prompt in taking back any quality good of any amount if a consumer desires to return whatever be the reasons. They have a national spirit.

In fact now market surveys and consumer surveys are being floated to know public views. That's the way public participates. A heavy amount is spent on that and the companies do get the needy materials as feedback.

More so where companies do direct dealings with public

chances of quality degradation are less but where companies hand over change to other middle agents who are seasonal and not permanent then chances of quality going down is on higher side.

The measurements of the quantity too are part of quality. Sealed packages too can be opened and resealed. The products can be adulterated or changed that's why we find duplicate of most of theproducts inthe market and nobody can help. These duplicate sellers give a larger margin to retailers. The company is not aware and so gradually looses a business and reputation both. Even purchasing items from a standard shop does not guarantee the quality of an item.

Concluding it is to summarise as below:—

1. The quality of a product can remain good if the owner remains in touch with public.
2. Sufficient labels should be put on the item packs about its measurement, product no., capacity period, addresses of manufacturer etc.
3. The duplicate and such other goods sold must be strictly dealt by Authorities.
4. Competition is desired to be healthy and not negative. In this case Associations should come forward through a forum and decide as to code of action in that for a healthy environment.
5. The consumer spirit may be realised and goods not liked be accepted by paying the customer its cost as well as sufferings thereof.

14

The Poverty Alleviation Programmes and Environment

In the history of human settlements the craze for living in a better environment tempted human beings too conglomerate in groups; irrespective of the era whether it was the Kingdom of a great ruler or was at the time when the feudal lord system prevailed or was in an industrial era; adopting thereby the life-style, to which they were never accustomed—devoid of family ties, social customs and free living; from where emerged a new trend the physical movement of the masses, all over the globe; from small hemlets to the so called town areas.

The power carriers from the beginning knew well to enjoy and exploit a particular situation and so got trapped all migrants in the hands of a few, till time, and made conditions to topple the heads; through co-operative and other revolutionary movements; followed vigorously in Europe, which made everyone the beneficiary that set a very slow trend, by which eventually the communities improved and the latter reflected the degree of affluency of a country; that is more affluent nations showed the faster trends of development.

The American ghettoes still persist but with mostly new entrants. In poor countries no exit do we find of the inmates, as the generations opt for a set living environment and even follow heriditory professions for their livelihoods. The poverty, unemployment, social delinquency and other such social factors could be attributed for this. The reformists, though, have tried to bring changes but no headway has so far been reached in this direction. Unfortunately in the countries with democratic set ups these institutions acted as large vote banks for the leaders under whose tips, those people normally twisted.

The plight to the stage of their present housing stretches right from the time when they squat on the ground without roof to the stage of living under the thatched huts roofing and then switching over to the building of temporary structures with or without proper sanitary conditions where some times break out the dreadful diseases and epidemics that take toll of lives.

The studies done by WHO Red Cross, UNDP and other parallel Global bodies in the field with reference to Afro-Asian text is commendable. In Latin Americans and other countries still it takes hours to draw a handful bucket of water. A large campaign for improvement to health hygiene provisions of nutrious food to children and providing drinking water to all is launched up by them.

The joy, pleasure, amusements and sad plights for them start and end at poverty from where they hardly find any escape. The vision gets restricted as earning two times breads remains the life motto for them. The question of having a desire to store any thing does not arise. The latter in fact induces sluggishness in their attitudes. The children grow and develop themselves on their own seeing the allround environment for which the scope of the parents for intervention remains hardly any.

Its more convenient for the community to spend whatever they earn the very day. On no income day they may go without meals even. At places one who can provide two times meal to his family is taken as a successful person. At places few get the jobs of peons, watchmen, soldiers but beyond that it is exception. The unemployment conditions are compelling the youths to go for criminal offences and declaring those institutions as notorious and criminal breeding zones. What can't one find there? Everything like vulgur video films are shown there; the brothels run; the adulterated liquor is freely sold and the murderers are easily available.

The sites of such settlement are horrible, along the drains, swamps, on busy roads and highways, just below the air funnel belt near aerodromes, along railway lines or elsewhere where intensive noise pollution prevails, they live there. In the recent past a number of such areas were gutted in fire where besides a heavy loss of belongings, there was loss of lives too. Sometimes this is due to sabotation by the political parties to build or destroy the vote banks in their favour. Otherwise short-circuiting of electricity wires is its common cause. After the fire the concern is to none as the police after investigations holds no one guilty.

The attitude of the police since ages, perhaps could never be changed despite there being brought so many changes-social and economic in the society. There is hardly any value of honesty and justice. For ever they will remain as dead wood. Exploitation and then seizing monetary gains practice have given confidence and protection to many illegal professionalists. Hardly proper and timely FIRs are got lodged. If not as foes, than the police have not acted as friends too of the public.

The slum situation is now a matter of an international concern. On one hand the various kinds of monetary aids provided are falling, are meagre and above all, daily thousands of slum settlements are building up on account of political patronage due to continuous refugees influx. Though, negotiations and mutual settlements are sorted out between governments but till that time one can well understand the treatment given to the refugees. Awful !

From the appearance the thatch huts look horrible as gradually all of them become blackish in colour. The dust accumulates allround and with little water the environment becomes marshy. The sanitation as usual is bad. The presentation of males, females and children is awful. Religiously the residents normally have few temples, mosques or even street gods to worship. The school facilities may be there but few avail of it on account of some greeds, like for example noon meal served in Coimbatore schools attracts students to come over there. The facilities and basic infrastructure is poor. At Phillipines and China and say in every country to draw a bucket full of water there are huge queues. The stealing of electricity by connecting wires to the nearest municipal electric pole is a common phenomenon.

The future of present day slums and of those areas which are going to be the slums of tomorrow is of course a matter of concern. The efforts are warranted at least to improve existing environment which may be a beautiful habitat area of future. The existing slums have chances of improvement. Of course happening of some miracle for eradication of slums has always been a possibility.

In India a private sanitary organisation of Sulabh-Shouchalaya that provides easy community latrines and baths is being accepted as a result of which sanitation conditions are improving. The only demerit of this is that the permanent blocks built near by the slum areas remain even after the slums are removed or else they come in the way of the future development schemes.

For slum improvement providing monetary aids only would not help because when that would pass through various hands the misuse is obvious. So indirect systems that could be dealt voluntarily or with self help could be of more help to those people.

The Mohan's Concept (Mohan Vichar) argues that slums could be changed over-night if the presentation of the masses is improved. The presentation in the sense that let their dress, hair styles, behaviour be changed for better. Let the dry cleaning and beauty saloon shops be opened for them and be made free of cost. Let every individual be guided about his personality improvements from such ends. The nuns may run nursery, primary schools and let the children be well dressed with bows, ties, belts, short and shirt, of course given free to them and let them in turn teach their parents and family members to improve their livings for a better environment. Such sorts of experiments are highly desirable.

Man's love for biotic elements and animals is always there irrespective of the situation where he is placed. In case, says, Mohan's Concept, if we could introduce such elements in settlements, the environment is bound to improve and grow healthier. A Model Plan of slum areas incorporating such elements be prepared so that future slums could be the better living areas.

Sir Lewis Mumford, a renowned sociologist who worked extensively for the betterment of the slum areas and gave the doctrine—folk, work and place must be related ; would ever be remembered. The folk means the people, work means job place and place indicates the residence. He moved from one area to another, exhibited drawings and models in slum areas and that had a great impact in slum improvement. India needs such practice badly.

Urban renewal concept in UK and other countries has much attracted public attention. In UK full-fledged departments of Urban renewal run in every state under the centre. Its need was more felt as the health of residents suffered due to ill planned old dark houses having no cross ventilation and amenities. Prior to 1979 act, a huge grant was provided for the betterment of the houses, specially of war period terrace-housing. One of the important features of urban renewal is "Enveloping" which means improvement of the front elevation of the building which is done in groups with group-aid from the Government. This enhances the appearance and life of the building. Urban renewal includes providing missing amenities and even looks for tenants welfare. The NGO's and voluntary organisations have made such schemes successful over there. In India the concept of urban renewal is yet to be vitalised. The co-operative movements experiments have been successful to an extent but not that much. In UK the demolition and then re-planning of an area is done but for a small job and very little, as people are scared about it. The local Authorities however, give temporary shelter to the residents till their occupation of new houses. In India such redevelopment measures are treated as harsh and are not favoured. The conservation techniques are welcomed. The rehabilitative measures are successfully operated by the various development Authorities in India. The resettlement colonies are built where live millions of people. The EWS and low income group housing projects are taken for the people of weaker sections. Mr. Laurie Baker has done a commendable work in this field for the HUDCO group in India using local materials and construction techniques.

The slum laws are very tough and courts are busy with such cases, sorting out the disputes of landlords and tenants. The low rents compell the landlords to ignore the maintenance of the property and the tenants find it as just wastage of money to get the property maintained. For the lawyers it's a good business and they remain busy with their clients for years with little results. The urban land and ceiling act, apartment act, benami bill, arbitration are some other legal issues that require legal care.

The public is hardly bothered for the implementation of

any slum project. It's time when these people should understand good and bad about them rather than Authority's one way decision of implementation of a scheme and spending money just for showing efficiency with little utility for the users.

The role of various professionals, sociologists, psychologists, artists, architects planners and economist to improve the areas is important to help the human settlements. The old, ladies and children require maximum care. The poverty and unemployment is to be removed by the professionals by direct and indirect techniques. Let the flow of art, culture and music create a fine poetry inside the slums. Let the people be made busy in them so that from the evil temptations an exit could be made. The art galleries; theatres, clubs, zoos, museums, parks, etc. could be built near-by for recreation purposes.

A large net work of monitoring and check programmes provides good efficiency in the work. The slum projects should not be of very long duration say three months time is quite sufficient; otherwise what happens is that sluggishness results as officers get transferred, the Authorities change as a result of which the tempo of work too suffers.

15

The Future of Waste Management Practices for Energy Recovery

Ever since the dawn of civilization, the various human settlements, in a conventional sense, forcibly followed a set pattern of living by adopting recycling techniques from a micro unit to a large one which later, when analysed scientifically revealed valuable results. The management of energy recovery from urban wastes is one of them.

It is doubtful at the present time to achieve hundred percent success in the mission until present systems of collection of garbage and waste are modified with radical changes. Whereas the individual household is being looked up by private garbage collectors, the municipalities and public bodies are responsible for urban garbage removal. Until privatisation is upgraded and brought up at the municipal level not much success is in sight. The voluntary organisation can play an effective role in that.

Thus all remain sterile until the public participates and takes interest. An awareness among the people is to be created to arouse their interests and they may be informed about the merits; number one regarding doing away for pollution sake and secondly making energy cheaper, economical and more freely available in the market.

The cost of technologies has to be such that may generate and provide employment to the locals and generate enough to run the plants and for service to the people. This thus meets the social goals of the society.

At a time when the entire globe stands at the crusade of watching exploding population with need of energy resources and at the same time on which not even little dent could be attained the situation gets more alarming and serious. When more cities are sprawling out of rural canvas where large scale migrations are taking place, the garbage collection not only is orienting to the tune of causing pollution and epidemics but more so paralysing the life of the city dwellers. These situations are more seriously being felt by the countries which are surrounded by sea and have little land for garbage collection. The sea becomes the area as garbage collector that is spoiling beautiful beaches of all the world over and creating diseases thereby killing fishes and harming sea animals. Identically garbage put on land is causing permanent pollution and attracting birds harmful for the air traffic.

These energies require to be generated in every lane and in every street; strongly feel the environment leaders, as scarcity of such item must remain. It is hightime when planners, architects, sociologists, economists all may meet and plan for cities of optimum population say one lakh and join them together as that develops. There is need to develop a system for that. Not only this but water, electricity, food all require that orientation says Mohan's concept; a concept devised by the author and which is mentioned in his authored books on environment.

Do you think that roasting the wastes by infra red heat or by other sources as is being done at California Silicon Valley, or by burying wastes are signs of achievement? Moreover would that not require extra energy for burning?

Even few countries are exporting garbage to others and paying money. Says, Mohan's concept when you can manage well and upgrade yourself then why you do not peep inside yourself and realise what blunders the humanity is making which can be solved in minutes. If the people participate and approve such jobs extracting energy by simple management practices of wastes then really a bright future is in sight.

16

Energy Conservation in a Developing Country

Since the birth of human settlements in the world, the sense of requirements, spending and savings whether that was a food item, finances, fuel or some other valuable needful thing, remained identical to the present era. Moreover among communities a status difference marginal or superfluous, always persisted due to the environmental or by self efforts of the communities. No doubt a mad race since though started immitating the higher status settlement in terms of business and other fiscal issues but none bothered for the short comings of the superiors in terms of environment, which in case if was bothered then the immitators would have dual advantage, firstly to superpass the better status community in terms of monitory value and secondly could introduce measures of overcoming the would be sufferings.

Eminent CHIPKO Leader Mr. Sunder Lal Bahuguna said that conservation is such an issue that requires to be handled even at a household colony level and if not cared would become a parasite for not only the country but for the entire globe. An anciety city of Mohanjodaro, for a population of one lakh people, had all amenities like granaries for food storage that was predominant over all development.

Identically a need for energy had arisen for all settlements rural and urban and gradually its scope is spreading and its demand has also arisen. Whereas in rural areas, the energy diversion is towards agricultural growth; in the urban areas energy is required far industries for providing economic base to a town or a country. These industries are provided energies at such rates as are affordable but the extravagency in energy spending has not at all remained as a controlable item. In the private sector though this is still comparatively very well looked into put in public sector this remains totally ignored. Many agencies are under going heavy losses and eventually energy is being totally wasted.

In the recent times besides natural energies like solar power, wind power, hydal power, petroleum the technology has much more advanced. Despite the government's indirectly trying to control the consumption by increasing the rates yet the consumption of energy is increasing exhorbitantly. A bulk of energy products are being imported and many are subsidised by the Government agencies.

The socio economic pattern of the society is now purely energy oriented. There is though an awareness being created of natural energies like solar power but even then there is no improvement. The chulah gas has become a common feature of the household though in rural area Gobar gas is being used at many places. The wind mills and wind power too is replacing petroleum products.

Most of the energy so far consumed is extracted out of non-cyclable processes. However, now recycling of waste products and other systems are being followed for energy renewable programme. For that a public awareness isdesired and entrepreneurs are required to be given incentives to work on such issues.

A total privatisation for energy conservation can not be ruled out. Because then that would not socially meet the needs of the society as lower status people would not be able to afford. That's why most of the public sector undertakings are made incharge of distribution of such systems.

The public and Authorities play a great part in energy distribution. The public consciousness and participation is earnestly desired to determine what they want and what they don't. The medias role is effective in influencing such systems for energy conservation. By means of pamphlets distribution such goals could be met. The Authorities too desire to run the organisation by meeting with the public effectively.

Specially more result oriented jobs of the country where priority is needed; less priority jobs could be made to run slow by introducing direct and indirect systems by the authorities from time to time. Curtailment in consumption is desirable.

The population measures be properly checked up and a regional plan related to national plan be prepared for the future consumption of the energy and thereby envisaging suitable energy saving schemes and delineating conservation areas. Regular monitoring is desired for this energy protection.

An audit report should clarify the area where extravagency has really prevailed and the guilty zones may be taken to task through courts by serving legal notices in the interest of the nation. The Supreme Court and other lower and high courts role in recent times has remained commendable in dictating what perhaps experts could not exercise from the fear of Authorities and Politicians.

The energy conserved is a sign of effluency of a nation. Japan and other countries have risen to heights by energy controls. The developing countries may immitate the advanced countries and see how more priority jobs could be fed with energy and how maximum energy could be received and stored. Maximum production and less spending will be providing self confidence to a nation which may find even healthy to export that speciality and may use it at times of need. This may help the other needy countries. At national level this programme need to be worked out.

Energy produced or consumed should create least pollution. The increasing number of vehicles that use petrol be replaced with mass transits. A time has come when environment needs protection. These energies have indirectly affected the environment in terms of degradation. Indirect measures are warranted to save the environment. Much dependability is desired in solar energy and others involved from nature that would create no harm to the environment particularly the bio-sphere.

17

The Management Evaluation of Marine Pollution

By Rajiv Bhatt

The oceans cover 70.8% of the earth's total surface area and hold 1445 million cubic kms. of water. The sea water despite its abundance is not directly useful to man, because it's not potable. However it's a virtually infinite source of nutrition and minerals. Man's meddling with nature, chemical effluents, nuclear waste, acid rain and ever increasing presence of carbon-dioxide have resulted in the greatest ecological crisis known to mankind. The marine environment has been equally abused by man. Since ancient times the oceans are being used as highways by merchant ships to reach far-flung places. With the discovery of petroleum and the advent of oil carrying tanker-ships the rate of marine pollution has substantially magnified. Frequent oil spills can potently destroy marine ecological balance, and what we destroy we cannot replace, nor can nature out at this speed.

The introduction by man, directly or indirectly of substances or energy into the marine environment is termed as marine pollution. This can be in the form of waste gases expelled from ships engine after combustion, garbage, sewage, major oil spills caused by tanker accidents and deliberate discharge of oily water by ships. Pollution caused by tanker accidents and non tanker accidents may attract wide attention, but the effect of such pollution though massive is realised. Operational pollution from tankers is worldwide and a regular feature from tankers whilst tank cleaning and change of ballast, and therefore, is the major cause of oil pollution from ships.

Most crude oils contain wax/asphalt and other materials in solution together with sediments which may settle out during the voyage and form a part of the residue with any cargo remaining on board after discharge which may be about 0.2% to 0.5% of the cargo carried. If during the course of tank washing this residue is discharged into the sea in heavy concentrations, it will remain on the sea surface for a long time.

To check marine pollution (MARPOL) stricter regulations are enforced from time to time by International Marine Organisation (IMO) which is the maritime arm of the United Nations (UN). The effluent discharge conditions laid down in the MARPOL regulations prevent heavy concentrations of oil from ships being discharged into the sea and also limits the amount of oil, being discharged into the sea from a tanker. However, it will require a concentrated and conscious effort from ship staff and the shipping companies to adhere to the MARPOL regulations.

18

The Protection of Wild Life

The wildlife has since been taken care of, and always it is projected that the numbers of tigers and elephants are dwindling on account of regular poacher activities over which the strict laws framed even remain silent alongwith non-interference of bureaucratic attitude and favouritism shown from the political circles. But feels the author that those issues persisted earlier also and so would remain for ever. The point to be seen is how to increase the animal number by efforts despite several casualities. It is not known when the NGOs or rich entrepreneurs would take up the issue and make those in authoritiy confident about the former's integrity in maintaining a balance of wild life.

Is it not a healthy sign that every one may have like a jewellery piece some wild life feature at his home as was in the past that Jamindars were having leopard skin hanging on the walls of their houses called havelis and simultaneously enough number of animals roamed around their free hunting lodges. The restrictions created in fact are making public more conscious and those wild life items are becoming rare commodities; very expensive, for which business entrepreneurs are making good amount of money.

A good number of animals could be raised by decentralising policies and encouraging recreational activities at door steps of each one. Tell me, why I should go 100 km or more away from my residence to see a national park. Can't those activities be made available next to my residence? How is it that feasible cannot be ruled out but possibility is there. Let mini zoos be spread over like parks being maintained for public under the directives of the authorities.

Why not the development authorities be given such a chance to develop such facilities and have rangers appointed in their offices as part of that? Why is it that only conservators or zoo authorities or states have the only responsibility of that? Till such measures are not adopted by the various local bodies; under one authority says the author, those activities are not bound to prosper. The specialization, however, could befrom the key authority. Everyday the poachers activities are increasing as there is scarcity of leather in the market for exports. Why not then the forest deptt. engage a full squad of police personnel and magistrates as one special wing rather on depending entirely on the other deptt. of Police for such help? Unless and until that would be done the efficiency is bound to suffer.

The present zoological gardens, are assets to the cities as otherwise in those areas, slums, unauthorised colonies or urban development would have prospered. The job under conservation is first of all, to demarcate immediately, for the entire country, such areas and take them up in the ready possessions and handover to developing agencies so as to serve indirectly, the future generation from pollution and intensive urbanisation.

At one time the Delhi ridge covering apprx. 7800 hectares area was proposed for building up a sarari area. Had that been earlier approved then unauthorised developments could have been avoided and flora and fauna could be retained. At least then the Supreme Court would not have come at the last moment to intervene and save such areas from degradation.

As a student in the early fifty the author enjoyed sitting on the skin of a deer, often been brought from his native village over which religious activities were offered. There happened to be holes in the skin of that due to the bullet marks. Gradually these vanished from the scene on account of strict bye-laws by which except poachers the common man stayed away from the scene.

Late N.N. Mohan, Meakin Breweries Ltd. was presented by some of his well wishers a python cub that he managed to keep in a large iron grilled cage and later handed over to zoo authorities; a practice which mostly was followed earlier in forest areas and other lonely places. The author's mother when she happened to be newly married bride and lived in the village, often the panthers roamed freely and attacked their calves and then she used to fight back and snatch the alive animal from the jaws of the panther; a spirit that prevailed among the residents to fight back the wild animals. Often now in the residences when the snakes appear, the practice is that either those are killed or trapped and handed over to the zoo authorities. The western photographers have been to the various remote areas and have snapped various thrilling activities of the animal world. Even a cat and mouse playing and than swallowing the mouse by cat is very much interesting to watch. Recently two bear cubs were purchased by a foreigner for a few rupees from a local resident at Agra which ultimately when reported to police the cubs were resorted to zoo authorities. It was told that bear cubs besides being used for sports abroad is used medically for its organs.

The ecological conditions reveal that one depends on another and so removing any one from the scene is difficult. The Gir forest is the indication of saving the tiger species which was fast dwinding on account of hunters craze for tigers. Now most of the wild life institutes have switched on for computer system and things have been easier to think, talk, propose and implement. There is a talk of the management of wild life now-a-days.

The author remembers that how in Delhi during summers of 1958 when he with his friends visited the so called Delhi's Zoological garden then in the name of animals and birds only few peacocks and monkeys were only in sight. A few earthernware jars with water were kept for the visitors in the name of drinking water. From that day till now when we analyse, the only merit of the area considered is that it saved about 100 hectares of land from getting squatted in so called the fourth most polluted city of the world and at least a number of trees, monuments remained protected as a measure under formation of the zoo.

But in the name of the zoo, though, several conservators have occupied the management seats hardly any dent could be observed. The visitors though coming in large numbers, at the end of the day go back tired. Hardly any emphasis has been given for the phased planning and all activities are spread over the entire site, covering entire area that did not provide any link from one use to another and so remained that as an utter failure as the visitor remained lost all the time.

Don't you think for such an extra ordinary heritage the appointment of guides could be helpful and none are found on such sites though, on the monuments sites, busy with tourists, they are. At least says the author that we would like to learn from the site the life pattern behaviour and all about an animal or a bird.

All over the globe this awareness brought countries together in a close network with the result of which international conferences and acts were framed up to improve the existing scenario of wild life. The sanctuaries, parks, zoological gardens started getting priority in terms of the national and international developments.

The distinction of a place in wild life world is because of the existing varieties of the animals like Bison of Bindapur, Elephants of Periyar, Rhinoceros of Kaziranga, water birds of

Bharatpur, Musk deer of Kashmir and Garhwal Himalayas, the swamp deer of Kanha, Asiatic wild of Rann of Kutch, Asiatic Lions of Gir Forest, larger cats (tigers) of corbett Park, Ranthambore, Sariska, manas, Sunderbans, Melghat, Palaman and Bandipur.

From the ancient day hunting has remained to be a recreational activity and few had become professional hunters. Even during child-hood, teasing of animals gave pleasure to many. For Akbar the great it was said that he had in his times kept 3000 cheetahs exclusively for hunting and for his son Jahangir it is told that the later killed as many as 889 numbers of Nilgais during the hunting compaign. Even during British regime to please the English heads as many as 5000 birds of Bharatpur sanctury were slaughtered daily during their stay. Among the names of the most notable hunters are Mr. Jim Corbett, Mr. George Sanderson, Mr. Dumbar, Mr. Brander, Mr. Captain Games, Mr. Hume, Mr. Hodgson and Mr. Baker.

The hunting sanctuary aimed at hunting and meat whereas the wild life sanctuaries meant the conservation issues. The former rulers like that of Kerala, Mysore, Gujarat, Rajasthan, Punjab and Madhya Pradesh in India who enjoyed exclusive sovereignty over their states kept secluded areas for hunting some of which have now been turned as major wild life sanctuaries of the world ; Bharatpur wild life sanctuary, Sariksha National Park (Rajasthan) earlier hunting resort of Maharaja of Alwar, Ranthambore National Park of Maharaja of Jaipur and Keoladac Ghana Bird sanctuary of rulers of Bharatpur State. For Maharaja Sarguja (MP) it is told that he killed as many as 1100 tigers.

The legal buildings helped in the preservation of the wild life particularly with the 1972 wild life (Protection Act). Earlier to that in 1951 the province of Bombay approved the wild animals and wild birds act and in 1952 the Govt. of India constituted the Indian Board for wild life to advise the

Government. In the year 1958 the wild life preservation society of India emerged out. The 1991 act reconsidered the revision of Game Reserved as well as Closed Areas and banned hunting except as provided under Sections 11 and 12 of the act.

Compared to 1952 when there were 33 National Parks and wild life sanctuaries during 1963 there were 72; covered up under an area of 7700 sqkm. In this terminology four types of preservatories are announced. National Park, sanctuaries, games reserves and the closed areas. A sanctuary can be created by the orders of either the Chief Forest Conservator or by the Minister of the State Government whereas National Park can be created abolished or changed by an act of state legislation.

Among the reputed sanctuaries are Kaziranga in Assam which in 1906 was a forest reserve, in 1926 a sanctuary and in 1945 became wild life sanctuary; the Hailey National Park now known as Corbett National Park which was believed to be set up in 1936 and as the Nanda Devi National Park was notified in 1939 in Kumaon Hills of UP. The Manas wild life sanctuary in Assam was set up in 1905; the Mudu malai wild life sanctuary in Tamilnadu which came up in 1938 and Kanha Park originally a game reserve was declared sanctuary in 1939.

A brief account of wild life sanctuaries follow as below:

Year	National Parks	Sanctuaries	Covered area sq. km.
1970	10	127	25000
1980	Both combined as	221	Covering 2.7% of total geographic area.
1984	19	210	
1989	24	372	11 million hectares or 3.38 of total geographic area.
1991	70	412	4.5% of total geographical area.

Among the various ancient times the big cat, the tiger has remained to be the hero of the wild life scenario and so the hunting of tigers was very common a feature among early period rulers as well as the hunters, the craze of which made people go from one place to another, in the world. It was all of a sudden when in the public suddenly the poacher activities came under notice and dwindling species of the animals realisation came up. This forced legal acts to come into force like 1972 Wild Life Protection Act and of 1991 subsequently.

In the newspapers various stories of the authorities, poachers and public came up about the declining tiger population. But what can be done? All are helpless to control despite there being strict laws. So until the highest Apex Court intervenes and gives specific orders for protection the issues are bound to remain as those are now.

In one of the visit to the wild life Institute at Dehradun, it was found that they were working on integrating various tiger reserve areas by keeping animals away from the public. How far they would be successful, cannot be ruled out but all was set on the sattelite computer system called the remote sensing techniques.

Conservation

Like other systems the conservation plays a vital role in the field of wild life specially for the endangered species. Besides malpractices adopted by poachers and other entrepreneurs efforts are to be made for conservation which really meant for protection of wildlife. On Jan. 9, 1992 the Ministry of Environment and Forest prohibited to set up any industry within 10 km of tiger reserve and 25 km from the boundary of the Sariska National Park of Alwar Distt. This however, annoyed the local politicians.

There are several Wild Life Institutes to train and

educate people among which Wild Life Institute of India, Wild Life Research & Education Dehradun and Crocodile Breeding and Management Training Institute are important. It aims at opening of various courses and researchers as well as various consultancy project jobs are also assigned. The world wild life fund also showed concern for the conservation of various wild life species.

The world conservation surgery was launched in 1980 which was to take care of endangered species. Another strategy was followed in 1992. In 1993 wild life action plan floated which aimed to; stage a network of protected areas, its management, linking of protected areas, restoration of lost habitats and education as well as researchers in the field. It also bothers for an effective legislation coupled with conservative surgery and assistance of voluntary bodies.

The conservation of wild life awareness was aroused by Stockholm conference in Human Environment (1972). Mrs. Indira Gandhi made efforts for success of the conference which was followed by 1972 Act. Further, the Jambhoji Bishnoi code of faith was formulated under Jambhoji Maharaj (1451 - 1531) for love of nature against religious superstitions and cruelities. Such operations operate in Rajasthan. For Khejdali trees production, Mrs. Emrti Devi's role is supreme; in 1730 whose example was followed by 363 village women who were mercilessly slaughtered. The Chipko Andolan and Gaura Devi efforts to protect himalayan forest is also worth mention.

The national forest policies of 1952 and 1988 did little help for conserving wild life. On the Bio-diversity the Rio convention specially has stressed on Bio-diversity of living organism that include marine and other ecosystems. The Rio convention further stressed on 'ecosystem' of plant, animal and micro organism communities and their non living environment. On the Habitat issue it meant the place which could be land, water vegetation and natural home of wild lives. The protected area meant the area of national park and sanctuary, prohibiting human activity therein. The 'sustainable use' meant to use the needs of present and future generations.

Management

The management of wild life administrative work and practice is an urgent feature. National forest policy of 1952 helped in protection of animals which also stressed on launching of Project Tiger in 1973 and National Wild Life Action Plan of 1983. The Director of Wild Life preservation is a post appointed by the Central Govt. and under him are Directors. The Chief Wild Life warden is the head who makes the policies and programmes. It is the link post between the State Govt. and the outside world of wild life. The Act of 1972 throws responsibility on chief wild life warden to take decision for animals which became dangerous to human life or are diseased, to allow any one for hunting or for permitting to prepare medicines out of animals. The killing of an animal for self defence is not an offence but at the same time can not be condoned. He has power concerning protection of plants. He grants licences and has a right in land acquisition. The related powers of sanctuaries and of punishment concerning prevention are also worked out. The wild life advisory board constituted with NGOs and members interested in wild life, provide vigilance. Its members are the Minister Incharge of Forest Officer, Chief Wildlife Warden, Secretary of the State Government, two members of state legislation and others. The board advises the Government on various wild life issues. The Central Zoo Authority is the guardian for all the zoo animals and their welfare: The funding is to be accounted as for 1992 act. The recognition of zoo is also through Central Zoo Authority. The major areas are Kaziranga Simplipal, Palaman, Melghat and others.

Tiger Project

The tiger (Panthera Tigres Linnaeous) is the national animal of India. Among the Indian race Royal Bengal Tiger is the most prominent. In the 1972 census only 1872 tigers were marked. The launching of the Project Tiger was made in April 1973 by the World Wildlife Fund (WWF). The ICUN, International Union for Conservation of Natural Resources met in New Delhi in 1969 to protect tigers.

As many as nine tiger reserves were established during 1973-74 under pooled resources of central and the State Governments; Corbett (UP), Ranthambore (Rajasthan), Kanha (MP), Palaman (Bihar), Simplipal (Orissa), Sunderbans (West Bengal), Manas (Assam), Melghat (Maharashtra) and Bindapur (Karnataka) with core areas of 4248 sq.km having a tiger population of 268. As many as 10 new reserves are added. 3 more are proposed for 93-94; at Bandhavagarh National Park in MP, Tandoba National Park and J.L. Nehru National Park both in Maharashtra. Thus total would be 22 reserves covering 28017 sq.km and with core area 12634 sq.km.

Three major issues human exploitation; maintaining ecobalance and reserves in flora and fauna are taken care of. The Ranthanbore poaching was discussed in Delhi on 13.7.92 with the Ministry of Environment and Forests. The tiger population accounted to 268 in 1973, 854 under 15 reserves in 1984, 1327 in 17 reserves in 1989 and now it is about 4334.

The Economic Development Project of 1989 under UNDP programmes was accorded for fodder for animals and provided $ 213,000 for preserving bio-diversity around 6 tiger reserves including Gir lion sanctuary and Rajaji National Park. The cat species group regional conference of 1992 made remarkable concluding remarks. One of the important conference on tigers was held in New Delhi on 10-13 November 1992 by the International Union for Conservation of Nature. Another International Symposium was held in

1993. Also in 1993 there was Delhi Declaration on tigers for poaching, biotic pressure on population, no research being undertaken, more of tourism activities and with in protected areas mishappenings. The Corbett Park initially had 323.75 sq. km and in 1966 it rose to 523 sq.km.

The licenses are issued for special games; for big game hunting, small game hunting and for wild animal trophy. A list is there where right to hunt the animals is there except the endangered species. There is nothing like right to self defence as that is no general excuse to kill animals. The Wildlife (Protection) Act 1991 regulates direct penalities to be awarded to detained.

19

Environmental Upgradation Through Efficient Management of Water Resources in the National Capital Region

The existing water resources in the National Capital Region whether those are rivers or underground sources are insufficient to meet the future water needs of the areas covered under NCR. An exhaustive management to augment the existing water resources and to conserve other resources at village, town and city level is desired with the help of panchayats as well as by the public associations through the Government of NGOs active participation. The storage of rainwater is advocated to meet the challenge and the means are required to be sorted out to reduce the pollution of water. Besides, the river water fronts are needed to be attractive for active recreation for particularly weaker section of the society.

The water resources whether those are rivers, lakes, ponds, wells or springs, since the beginning of the history have remained to be the focus of all the human settlements and whenever such spots of water resources changed or disappeared through the natural course, the settlement pattern too altered and sometimes even vanished. The public baths of Mohenjodaro 3000 B.C., the Roman Thermaes and the practice of building Indian baulies are the past examplary indicators as prominent features of the past cities.

In the past for the environmental improvement the Honourable Supreme Court of India intervened particularly to protect the Delhi's Aravali ranges against encroachments and hazardous industries. Then recently the Supreme Court order dated 11-10-96 1A No. 29 writ petition no. 4677 restricted the construction/development in the area outside the green belt upto 1 Km radius of monsoon fed water bodies, Badkal Lake and Suraj Kund in Haryana excluding Delhi and likewise it is expected that in terms of protection of water resources it may take interest. Should we then expect it for betterment or otherwise? The W.W.F. India has also recently moved to Supreme Court for protection of Wild life.

The eminent personalities on environment Mr. Sunder Lal Bahuguna raised the hill people's voice to protect Shivalik hills from the ecological imbalances caused by the construction of Tehri Dam and identically Mrs. Medha Patkar spoke in favour of protection of Narmada river in Gujarat. However as per the latest position to raise Narmada dam to 110 m from 80.3 meters is being debated. Contrary to this the Hon. PM reference in his speech in ASSOCHAM meeting at Vigyan Bhavan on 16th Oct. 1996 highlighted the issue that whenever any such mega project is being worked out such environmental issues are taken up as if the projects would be destroying ecological impacts. Then, how come the country's employment and other problems is going to be sorted out?

The Yamuna River Front

The planning of Yamuna River fronts, attempted from time to time appears to have been half heartedly dealt without examining the physical issues. The author published an article in Design incorporating Indian Builder Journal July-Sept. 1981 on river front under direct guidance of Shri M.N. Buch. Shall we appreciate a river bed landscape with green lawns and trees or a sandy bed with melon crops? The latter is definitely desirable. The river bed is marked under zone "O" in the Delhi master plan and ironically whenver any deficient facility in the adjoining trans-Yamuna area of Delhi or at other places are desired, all those are proposed on the river bed when that would be planned. Hence in advance those have been marked over there leaving little scope for it's proper development.

The 1914 sanctioned plan from NDMC of New Delhi by Lutyen's shows a large reservoir in front of Purana Qila to feed the Yamuna water to Viceroy House, which some how did not materialise. Compared to that what we have done now? Developed all the river bed area adjacent to Pragati Maidan, for which Mr. Patwant Singh, a noted journalist in one of the architect's conference, felt much concern and still desired that to be converted as the river bed.

What do you think is the future of Yamuna river bed? By 2051 it is believed that the entire bed would be concreted from top to connect trans-Yamuna areas. When countrieslike UK and others are appreciating our river water spans, we are deadly bent upon spoiling those.

When the bridges near Inter State Bus Terminus, were at planning and construction stage none had a say, that those would be damaging the river side view and that is what those bridges have really done. To connect Inter State Bus Terminus, Delhi, a large amount was spent. May I ask, was there no possibility of shifting ISBT? Then if that was so, would not we have been saving crores of rupees?

Trans-Yamuna Settlements

The Noida & Delhi's trans Yamuna area are two major settlements of NCR. The trans-Yamuna Delhi constituting 8797 hectares and 3% of total NCR area would contain a major 27 lakhs population by 2001 which would be about 12% of total NCR and which would be a threat to the proper development of the water resources of NCR area. Though, it is fed with the Bhagirathi treatment plant along with Yamuna water, still 50% of it's population living in approved unauthorised colonies and urban villages are depending upon shallow hand pumps for drinking water. During 1988 the cholera broke out at Nand Nagri tolling heavy lives. Mohan's concept says that if at all there is a devastating flood then water would jump over the bandh areas and Kishan Nagar, Gandhi Nagar along with other areas of trans-Yamuna area would be swept down as the roofs of 4-storeyed houses are even lower than bandh line. The Sanjay lake at Mayur Vihar is the important water body of the area.

Channelisation of Yamuna River

For the river Yamuna, the proposals are floated like channelisation of the river to a width of 550 meter in its entire reach from Wazirabad to Okhla and developing of the river front. Though this river is the major source of the water for the NCR area, yet it has remained ironically just like as a simple drain. It's wider banks are savagically been occupied by unathorised encroachments. More over the authorities like CWC, Delhi administration, floods deptt. and others are anxiously waiting to put planned structures in the name of channelisation over there. The rich enterpreneurs too are eagerly waiting for a favourable decision. Of course this would generate huge flow of money and employment too would be generated but at whose cost? The only open space available for Delhi would vanish and the water level of the river is bound to rise. Before doing any such experiment it's consequences must be forecasted as have been experienced in Victoria Lake of Africa.

The NCR shall cater to the needs of the population of 128 lakhs of UT of Delhi and 49 lakhs of other states in NCR by 2001 for an area covering 30,242 sq.km of states of Uttar

Pradesh, Haryana, Rajasthan and UT Delhi. It would constitute 6677 rural and 94 urban areas. Total NCR population in 2001 would be in tune of 234 lakhs with Delhi's 129; Haryana's 38; UP's 63.5 and Rajasthan's 3.5.

To serve the water needs the major water resources of NCR are; river Yamuna (Major source), western Yamuna canal, Eastern Yamuna canal, Agra canal, Gurgaon canal, Hindon cut and Sahibi river.

The Canals Network

The fate of eastern and western yamuna canals which start from the right bank of river Yamuna at Tajewala have remained most unfortunate. Imagine the talents of our ancestors Feroz Shah II, who planned western Yamuna canal in 1356 and Mohammad Shah 1718-1748 who laid the foundation of eastern yamuna canal which got completed in 1854.

The western yamuna canal earleir was designed for a capacity of 396 cumecs of water and for irrigation of 7.12 lakh hectares land at Haryana. It had it's total length of 5505 Km and flowed down from north to west in Delhi. On it's bank particularly at Hyderpur, Pitampura, Shalimar Bagh and Ashok Vihar in past enormous orchards were located which with the time have vanished. The tall green grass called "Bhara" no more is to be seen. The water that flowed down to Gulabi Bagh and irrigated the cultivated land there has been cut short and falls now to Indra drain with some water being sent to nearby industries along Tri Nagar side. The most unfortunate part was that an area which had extreme possibilities of being developed as Picnic spot remained unexploited. Had the Supreme Court orders as applied to Badkal Lake been issued for this area earlier, the scenario would have altogether been different. The water of western canal further deep from Azad Market side was reused by Moguls in their planned Faiz canal, that flowed in the centre of Chandni Chowk in the walled city of Delhi. In fact, the great moguls had a taste and we do find placement of aquaducts on north of Red Fort and in other areas.

The Eastern Yamuna canal originally designed for 85 cumecs capacity and meant for irrigation for 1.91 hectares area in U.T. is 206 Km. in length. Flowing at it's last end in the trans yamuna area of Delhi it ends abruptly near G.T. Road. At Delhi it is treated as no better than a mere drain. At one time while planning Dwarka scheme in Delhi proposals were to transport water from westernYamuna canal, but since as already sizeable population grew at Pitampura and in vicinity, there was no chance left for that and the capacity of Hyderpur treatment plant was increased to meet the local needs of water.

The Agra canal at U.P. takes off from the Okha weirs and is 160 Km long. During pre monsoon period it's flow is supplemented from Hindon and by upper Ganga canal via Jani escape. It irrigates 1.38 lakh hectare area in UP. This canal however, remains most polluted. The Gurgaon canal carries a discharge of 63.53 cumecs. It is created by remodelling Agra canal. In it's 8Km reach from its off take at Okhla weir, the Hindon cut covers vast area around Ghaziabad where maximum aqua vegetation is done. It rises from Shivalik hills in Saharanpur and after traversing 150 Km. enters Meerut. It meets ultimately Yamuna downstream at Okhla. A number of encroachments have come up near Kondli Gharoli area in trans-Yamuna area that a proposal of 45 meter road development along it at Delhi is lying pending since long.

There are three major irrigation systems of Ganga canals:

1. Upper Ganga canal,
2. Lower Ganga canal,
3. Madhya Ganga canal.

The lower ganga canal is fed from water works at Narora (Below Bhim Godwa in Bulandshahar distt.) and is 1160 meter long. The Ganga water is treated at Bhagirathi treatment plant at Shahdara. The visit at the site revealed that a very high chlorination doze is injected on the plea that by the time the water reaches far flung areas the doze would be OK but what about the near by where water is consumed at once ?

A visit in the past was made to Dhansa Bandh and the area previously called Najafgarh Jheel whose reference is found in the history of Delhi, surprisingly, there was no such lake and all the adjoining catchment areas were either encroachment or were being cultivated partly with aqua vegetation. Imagine a large water shed vanishing before the eyes of Delhites and not a single voice was raised to restore that. Had that been functioning, says Mohan's concept there would have been no water problem in Dwarka scheme of Delhi which shall have to depend now on entirely ground water. But how long? Can a million population survive on such resources? However, on paper it is estimated to work for 10 years and then afterwards would be seen. The Sahibi river that is connected with Dwarka scheme starts from Bairath in Jaipur and after traversing 145 Km in Rajasthan, enters Haryana and further Delhi through Dhansa Bandh first discharging it's water into Najafgarh Jheel and then ultimately to river Yamuna through Najafgarh drain. Even the Jheel existing now would no more remain when population would appear and so would be converted to drain. The Jheel and najafgarh drain are located side by side at Najafgarh; Jheel on one side and drain on other, the two meeting and a little distance and forming Najafgarh drain. After 1978 flood a Master Plan of Sahibi River was prepared under Central Water Commission to augment; the construction of Barrage near Masani, the renovation of Najafgarh Jheel the construction of supplementary drain to regulate the flow of Sahibi river to Yamuna river.

Ground Water

The old residents of Delhi must be remembering the Delhi of 1950's when on account of high sub soil level the tube wells were ordered to be dug up all over the urban areas to decant off the soil water level as mostly the water penetrated out in the ground floor houses, damaging thereby household activities. Surprisingly, from that time now we have reached a stage of hollowing up of our sub soil level for which extra discharging of water is required to retain the standard water level. Probably that time it was all government accommodation with regular water supply but now due to excessive urban growth and lack of regular water supply mostly public depended on underground water resources that shallowed the ground water. But the question is how to manage the water necessary for the settlements at regional level covering 30242 sq.kms area with 6677 villages including 4 urban areas. Due to extensive pressure water table remain declined from 2 M to 8 M at Delhi. The Delhi soil is off Alwar quartzites and rocks of Aravali mountain as well as alluvial plains on east and west of the ridge. The climate is semiarid; average annual rainfall is 675 mm, the chemical quality of water is brackish to saline at shallow depth. The higher concentration of fluoride too are detected.

Pollution

The pollution of yamuna is due to industries, agriculture, animal husbandry and on account of domestic use of the Industries in Meerut, Bulandshahar (UP), Karnal, Sonepat, Rohtak, Gurgaon (Haryana) and Alwar (Rajasthan). The Hindon contributes a great deal of pollution in the river on account of organic pollution and nutrients. About 70 per cent of the inland waters are branded as doubtful for human consumption. More than two million people due to water borne diseases and over 50 million people are incapacitated annually. Delhi is a paradise of the wide spread breeding of

Typhoid, paratyphoid, dysentry, cholera, haepatitis, gastroenterities and many more. About 50% of raw sewage flows into Yamuna. The Najafgarh drain & DDT plants create greater pollution. The frequent power break downs bring the sewage treatment tanks to a stand still resulting in dense sewage going to river. The statement made by Professor Arceiwala that "Agra may be drinking Delhi's sewage" is to an extent true. During 1977 jaundice broke out in Janakpuri, Delhi.

Drainage

The main drainage of the area under NCR flows through Aravali ridges and the Najafgarh drain. At Delhi; Aravalis which is the main drainage source covering an area of 7777 hectare in Delhi forms a large chain of hills starting from Surajkund going towards north to Tuglakabad, turning west to Qutab, Naraina, Anand Parbat and further traversing inside towards New Delhi at Budha Garden, Idgah, Hindu Rao Hospital, Rajpur Road and ultimately ending at Timarpur touching the river bed. The author earlier proposed the joining of the ridge to the river from the front of Jubilee Hall of Delhi University through Khyber Pass across Mahatma Gandhi Road near Majnu ka Tila area. The Green Party of India subsequently agreed to form a human chain at that time and Sh. S.L. Bahuguna also appreciated the proposals. We talk and see in a straight row and do not desire to take pain to develop a diversity. However, this factor is well looked up by foreign experts. If there is a river, let it be joined to city parts; the ridge and urban areas through artificial surgery. What is the harm? Says Mohan's concept that if the Aravalis and Yamuna join hands near Khyber pass there would be found every day at least one lakh persons going there for recreational needs which do not exist at all in the north west Delhi areas.

Floods

The flooding of by rivers has been a common phenomenon paralysing the lives of the people for months and tolling heavy lives. The floods and disaster management is looking up such issues. Floods in Yamuna normally appear during August and September months. The devastating flood of 1924 and 1978 were on account of heavy rainfall in catchment areas and depression in Bay of Bengal, whereas 1947 floods was due to depression in Arabian sea. The 1988 flood was again a repetition of heavy rainfall in upper catchment areas and low pressure in south east Rajasthan. Even areas that did not earlier get flood water started receiving the same on account of creation of bunds, deforestation and urbanisation. The geographical area of NCT of Delhi is 148,400 hectare. and the maximum flooded area in a single year from 1953-1984 is 70,000 hectare whereas the total area flooded upto March 85 is 84,000 hectare. Consequently for a better discharge the Najafgarh drain is remodelled to the capacity of 253 cumecs.

A proper water shed management is required earnestly to check floods by growing forests, structural works and grass growing. The flood forecasting, relief work and emergency operations require proper monitoring.

Conservation

The conservation strategies for water is required in terms of storage of rainy water and flood water by long-term means by dams; like Tehri Dam on Ganga at Tehri, Kishau dam at Yamuna Dak Pathar and Renuka dam at river Giri Renuka project. The dams embankments required proper maintenance for the water shed management. The storage is required at all levels by proper management systems which require good finances. The system dynamics is one of the important factors that require adequate care. Other components of this are storage; natural and man made like rural bandh, historical water bodies and village tanks. Natural drainage channels are Bhurianala, Najafgarh and Indori Nadi in NCR. The Delhi master plan could however, look far harvesting rain water at Wazirabad left and right banks, horse shoe lake at Bhalswa,

Burari, Kalindi Kunj, right bank of flood plain, Nala Mandela, Tilpat Tuglakabad Jheel, the Bhatti mines, Tejpur querries which could be used as areas for storing water.

The Development of Eco parks of Rohini drain, Wazirabad and Noida would or would not help can not be ruled out but at least this may raise the water table and enrich the bio-diversity. The demerit of this is that in urban places those occupy large areas (25 hectare). Hardly, any efforts have been made for restoration of the lost Najafgarh Jheel and it's catchment area as well as areas of old Ganga bed on Hindon at Ghaziabad, the Hauz-i-Khas lake, Anangtal Hauz-e-Shamsi, Dada-Ji-ka-Hauz. There is no proper maintenance, repairs and cleaning of dams and existing bands. The ponds with silting problems require proper management. The fresh water to be totally banned for toilets can not be ruled out but it is not necessary. Also attitude of the authorities to spread land fill sites any where and everywhere require approval from water authorities as those may not be close to the water bodies. The depleting forest cover all over ridge shall be restored; is doubtful seeing the present urbanisation. Any way efforts could be made in the region to create forest reservoirs on ridge (Masudpur area) andothers for water protection. The artificial recharges along with refilling of acquifiers along flood plains of Yamuna during monsoons are recommended. The cropping pattern requires to be changed to save available surface water for drinking usage. The scheme to use saline ground water be thought of and no landfill site should be near saline ground water. The laws require adequate checks.

Legal Aspects

The legal issues related to ground water be looked into. One had a right over the ground water lying below ones land as per Section 17 of the Basement Act; 1982 and Section 3 of the Transfer of Property Act 1882 read with Land Acquisition Act 1894.

A person who transferred his land also transferred the water under it. The four agencies which are looking into such aspects are:- (1) National water policy for integrated and coordinated development of surface water and ground water. (2) The Tamil Nadu water policy 1894 for action plan and recharging of ground water. (3) The Model Bill ground water (Control and Regulation Act 1970 for regulated ground water extraction). (4) Bombay Irrigation (Gujrat Amendment) Act 1976. This brought in a system of licensing and boring below 45 feet depth requiring a license. It also stated that water can be used for drinking or for agriculture.

Delhi's Situation

Delhi requires special attention on account of growing population of 122 lakhs of Delhi as per MPD 2001 in terms of its water needs. To meet a gap of demand and supply of 12.80 mild is a hard task which could hardly be covered up by Hathnikund Barrage. Ravi Beas System, Sahibi river, Okhla weir, Bhagirathi plants, Satluj-Yamuna link or by the basins of the adjoining states. The Delhi's short fall in water consumption is expected to be 15% by 2001 to be met by Ramganga water. Out of 504.90 MCM of ground water about 29% is short fall, not fit for drinking. The 45.1 pcd non-domestic water (industrial and commercial water) in a year comes to 184.73 MCM that needs treatment at source and not to be sent to river back. Usually 50% of river water is for domestic use and return goes to yamuna. More treatment plants are required for Delhi and the existing Chandrawal Wazirabad, Ramganga water diverted to Bhagrathi plant, Ravi Beas water treated at Haider Pur treatment plant are not sufficient.

The yamuna river board constituted for coordinated management of Yamuna upto Okhla to be activated in tune with NCR systems. The Ministry of Water Resources, Govt. of India resolution of 11 March 1995 for construction of dams Kishau Lakhwar, Byasi project, Hathikund barrage and identifying Chatra, Chami, Arangpur, Dhakot storage projects require to be well monitored. The national water policy of 1983 approved by national water resoures council giving states the power of water and centre of review river valleys require to be updated. The standard formed by 8th plan for water consumption of 125 lpcd for places where sewers are existing and 40 ipcd otherwise as well as 363 Ipcd recommended for Delhi are less which require adequate modifications subsequently.

Recommendations

- It is felt very strongly that the head of NCR body must visit each and every small source of water and dictate the development as per various categories of developments envisaged.
- Mr. Sunder Lal Bahuguna rightly said to start with a strategy. Let every residential lane have a rainy water reservoir.
- Just like Surajkund complex, no structure or development be allowed to come within 1 Km range of any such major water body of proper lake or canal level.
- Inviting Supreme Court for deciding the fate of water bodies at this level is earnestly desired to avoid future crisis for wells, springs and johads. No structure be allowed to come upto two meter distance from smaller water sources and identically for river and lakes adequate measures be set up. This is a must says Mohan's concept for healthy development of NCR.
- Even linking of all smaller bodies is desired at all levels.
- Many ambitious plans are chalked out for Yamuna river, like channelisation and water front developments

but says Mohan's concept that let the river be left alone and remain as it is and leave it for it's own destiny otherwise the consequences could be adverse even, for which the city people must remain prepared for.

- The grass root systems need polishing and from a small spring to a river, the media may come up to involve public by specific programmes of construction designing and funding to local panchayats.
- The attitudes of public need to be changed and sanctity to worship water may be aroused. The sufficient training programmes in this direction need be worked out.
- The technique of remote sensing would help delineating all water resources of NCR as well tying outside that. A master plan for those could be proposed for the development of water resources.
- Would all the planning at regional level benefit the masses or aggravagate their problems in future is to be seen right now. In that case regional plans would require more checks, controls and monitoring or even scrapping if at all that is not practical.
- While doing the entire exercise the social parameters must be well looked up about any beneficial programme.
- The copying idea of any reservoir, dam that has worked well could be taken as a guide.
- The beautificiation of drains may be taken up in near future for cities. Even now the few drains at Delhi have been concreted from top and are being well used.
- The studies beyond region even are required like for

example receding of Gangotri glacier that would affect agricultural output by 2003 ; as studied recently by International Commission of Snow & Ice.

- The stealing of water and mixing of sewage in drinking water requires proper checks.
- The clubbing of environment health and urban issues are a must for a healthy development.
- The sufficient schemes relating to water be chalked out to generate employment for which required finance need to be augmented.
- For the conservation of water bodies sufficient efforts be made for the system dynamics for water storage, harvesting of rain water, building reservoirs along rivers development of eco parks, afforestation measures and banning fresh water for toilets.
- The National water policy 1944, Tamil Nadu water policy 1944, the Model Bill (Ground Water Regulation Act 1970), Bombay irrigation (Gujrat Amendment Act 1976) all require co-ordination and amalgamation in to one single bye-laws management systems.

Yamuna River Front Recommendations

Exclusively for the river Yamuna front few recommendations are being mentioned as below. However, for details study from Wazirabad to Okhla as mentioned earlier an article written by the author in Design Incorporating, Indian Builder July-Sept. 1981 may be referred. A totality concept of the river bed is desired to be reflected in terms of the relationship of the river flowing further deep towards north and south. For creation of linkage of the Delhi ridge and river Yamuna despite the former being the most dominating environmental feature of the city the traces of which are still found in the river beds near Majnu ka Tila area to the Suraj Kund bed at Tughlakabad are required. Shall we get a landscape environment of trees of river bed or a river scape with sandy dunes and melon crops which would be highly appreciated? Ideally a river scape looks natural with sandy beds. However, for the flood prone areas plantation could be done. For the provision of the various samadhies, memorials and future provision of that need to be given due consideration.

A clear cut picture of the delineation of the developed zones for the migratory birds is required as no study has yet been made as to where and at which place they are presently concentrating. At the sites of village Usmanpur and Wazirabad such birds are found in the riverbed. Near Krishan Nagar at Rao Tula Ram Marg and at other strategic locations from where the river bed looks extremely beautiful, such spots require identification as from that location, at least one gets a silhouette view of the Red Fort as well as of the trains running above old steel bridge alongwith a site of washermen drying long size yarns that look beautiful. Mr. M.N. Buch liked washermen drying long cloth yarns and desired to have the same at Jagatpur and deep north, where there are beautiful locations on the river bed which can be developed. Should we really go for building structures over the river bed, the only left out open space available for Delhi; the 4th most polluted city of the world? Obviously it would be made worse by the developers themselves.

The efforts could be made to relate this river with the other water bodies like Eastern Yamuna canal and the development of ferry services along the river bed in either direction be thought of. The efforts be made to provide relaxation and recreation of the poor strata of society which dominates the trans-Yamuna urban scene as they really can't afford to pay for expensive recreation. Also, a parallel ring road in the trans-Yamuna area could be developed to facilitate traffic. A large number of Jhuggi Jhoperies have come up along the river bed for which there is no policy of resettlement and which is really required. The unauthorised approved Colonies on the river bed require to be linked up. The water sports activities, the locations and role of that specially for the poor and middle class societies is desired to be properly planned. A provision of storage of rain water in well for future provision for water supply for the city area be made. A programme is to be thought of introducing this river scape inside the city. The pollution aspect has to be sorted out in detail like in the past the cadmium contents from thermal slag of I.P. thermal power house went in to the river and those who consumed fishes, fell ill. The uses like drainage, sewerage, garbage, bio-gas, poultry, fish ponds, be avoided from being close to river.

The river front study has to get proper green signals from the flood deptts. for going ahead. Would the copying of Amsterdam and Thames River for developing river beds at Delhi, be helpful is doubtful. The linkages to river from the Jama Masjid side and as well as from the Ramlila ground side and from the Kudsia Park is desired. The public opinion in this case is required to be known about what the public actually desires for development of this zone. Would so many uses on the river bed not create health hazard in future has to be looked up by the health experts.

However, examining the past proposals the uses suggested like light serving industries, wholesale warehousing and cold storage, residential uses, sewerage treatment plants, solid waste landfill site, Govt. offices, institutional uses, administration offices, fire stations are to be reviewed properly whether we really desire to have such uses on this open belt. The race course, cycling pedestrian walkways and wind mills location could be tolerable planned uses along the river front. The land uses specified as parks, open spaces, Zoological gardens, huts, horticulture belts, nurseries, forest areas cremation and burial grounds can also go.

The North Delhi people specially have no recreational areas and as such the surroundings of Majnu ka Tilla area require exploitation where lakhs of people would go and enjoy morning walks or river bathing. The city requires a large amount of the wind from the river side on both sides, east and west so the wind directions are required to be properly kept as open air funnel areas. There is no thought of drainage, from the trans-Yamuna area, and Patparganj area that require to be linked to the river as they were in the past. The tourists entry to such areas appears to be doubtful as tourists like free environment, sandy beds, bathing areas, sun bathing and like that. However for the poor people, spots of sunsets viewing from the eastern side and the sunrise viewing from the western side could be developed with a natural organic growth, where lakhs of people could come and enjoy their mornings as well as evenings.

20

Energy Conservation in a Developing Country

To manage adequate energy for a country with a population of more than 90 crores is an impossible task, rather to call that as gygantical or leaving any scope to make that possible. Compared to the per capita energy available of a developing country, to developed, as 1/40th part and the day one may assume making both at par, ultimately can very well estimate of all the residual resources of a developing country vanishing over night.

Do you think energy is being saved now through sophistication in the equipments which are being made available in the market? Would those all repairing old machine shops vanish over night? Not the least. Then if those persist simultaneously then how can the energy be saved? Moreover, the less prosperous units have no alternative except to adopt repair job rather than to go for affluent sophistication that requires surplus money despite there being IDBI and other loan giving agencies.

Some voluntary agencies that have come up in UK may be successful in India too and units may start getting registered with them. Otherwise, efforts by Government or by individuals would be feeble unless such voluntary groups multiply.

When we say that 25% of the total import budget of the country is on oil over which we are depending more and more, then chances of the growth of population at such an alarming rate has been the matter of concern particularly in terms of energy conservation. As per March 1, 1991 census India's population was 894 million and about 26% of that lived in the urban areas.

The share of GDP in agriculture sector declined to 39.7% in 1970 to 29.5% in 1993 whereas in the Industries that has gone up from 17.4% to 22% during the same period. The average rate of growth of GDP which was 3.5 in 70's has been 5-6% during 80's and the trend continued.

The coefficient of elasticity of final Commercial Energy w.r.t. to GDP at constant prices has been 1.14 during 1970—93 whereas the elasticity coefficient of coal, electricity and refined petroleum products from 70—93 was 0.54, 1.83 and 1.43 respectively.

In 1970 the Primary Commercial Energy was 1841 kg/rupees at constant prices compared to 2445 kg/rupees in 1993 where the final commercial energy was 1441 kg/rupees in 1970 as compared to 1605 kg/rupees in 1970 as compared to 1605 kg/rupees in 1993.

The energy conservation efforts are not that intensive in India. The Department of Power in 1989 established the Energy Management Centre, an autonomous Organisation with foreign assistance from EEC, UNDP and World Bank. The National Productivity Council, Association of Chamber of Commerce and Industries (ASSOCHAM) & Federation of Indian Chamber of Commerce and Industries (FICCI) did a few research jobs in the field.

The dependence on the fossil sources of energy will increase 75% of over all energy demand as the oil consumption is already double to that of production. For energy promotion the technological innovations are desired to reduce energy usage. The Energy falls as the productivity gains have added to energy value used.

There are four major areas identified for energy conservation; industrial sector; transport sector, agricultural sector and domestic sector. Among these the most of the consumption of energy is only in the industrial sector. If at all any reduction on saving is desired, it could be possible only in this sector. There are three basic spheres to monitor energy; power generation, transmission and distribution sector. The least the flaws and leakages are there more efficiency is achievable. Now of course computers, softwares and database have come up and automatic machines have replaced labour jobs and so energy is bound to have better efficiency. A line action plan, fiscal incentives, legislative measures and a time frame work, short, medium and long term, can better orient the energy field.

The future scenario of energy is bleak as shortage of it is in sight. Those industries using energy 20-30% of the cost of production can be asked to reduce its energy quota. In 1989 the Energy Management Centre (EMC) started and both MOP & EMC identified several such key factors. In UK the most successful of all schemes is the voluntary agency, called Top Management Consultant Programme (TMCP). Those under TMCP have been benefitted by rich dividends and it could be well imagined that by 1994, 1800 such units signed for TMCP. India also needs such voluntary institute.

In the rural India fuel wood, animal wastes and agricultural residues are used to raise energy resources. The total energy consumption in 1992 was 16 $ of world coverage and that did not take into account the rural fuel woods and other resources used. The energy consumption KGOE per $ of GDP is highest in India 0.836 compared to .50 in S. Korea, .343 in USA and .132 in Japan. To reduce load on energy the alternative is to be thought of as a must in the form of a renewal resource of energy. The researches are desired by institutes like PCRA, Petroleum Conservation and Research Association group that met in 1978 for energy audit. PCRA launched schemes for loans, subsidies, technological upgradation in industries, transports, agriculture and domestic sphere for research, development, training and HRD programmes. The power utility operations, consumption norms need to be adopted in terms of attraction of tax benefits and export subsidy. The agriculture sector needs focus and adequate pricing of energy is desired at the end as suggested by many.

The coal is the biggest source of energy. It is observed that the role of commercial energy in the total energy is increasing. The coal is among primary sources of energy alongwith other items like oil, natural gas, hydro and nuclear power. The India's coal reserves are approximately 200 billion tons which is 1% of the total world reserves. The Indian crude oil and natural gas reserves are 80 billion tons and 73 billion tons which are going to last for another 20-25 years. The coal is meeting 60% of the commercial energy of the country. The coal based thermal capacity is projected to be 81,000 MW by 2001.

The energy consumption in coal is in two facets; one is source of energy and other consumer of energy. The improved exploration yields maximum coal recovery. Further the benefits are derived to improve quality. The utility of washery products is meant for boilers, brick burning etc. The effective utilisation of coal is required in sectors of power, cement and steel. The energy conservation in coal production could be taken up in phased programmes. The fire caused in coal mines is a sad affair; one at Dhanbad in India occupied 17 sq.km and the fire is burning there since last 70-80 years. The Coal bed methane (CBM) is an important part of coal or energy. The molten carbonate cell that is producing electricity from coal worth 60% efficiency, is being assisted by Japanese techniques NTPC and BHEL.

The petrol is a form of hydro carbons which is the main source energy. The maximum import of oil is of 25% total import bills, directs one to conserve it. The ONGC, Oil & Natural Gas Commission is an established Institute of Petroleum Safety and Environment Management (IPSEM) to Provide research and development in the field. The efforts are being made to conserve energy during drilling and to adopt heat recovery system. For off shore platform ONGC is using waste heat recovery for well fluid heating and glycol heating. Further jobs being done are to maintain ecological glance; doing cluster drilling and adopting social programmes i.e. construction of roads, drains etc.

The measure to curb pollution at drilling and production are required which is being initiated by Oil India Ltd. The Indian Oil Corporation is looking for energy conservation effects. With the setting up of a conservation cell in the Ministry of Petroleum and Natural Gas (MOPNG) energy conservation is being well looked into.

The refineries during VIII plan saved 25,000 tonnes of fuel per annum. In the name of new technology called PINCH technology which is for recovery of foreheat from hot streams is much advanced. The Madras Refineries Ltd. gave improved product yields.

ASSOCHAM has made various surveys of energy conservation from time to time and observed that whereas large industries were conscious of energy audit, medium and small were not. For energy conservatrion ASSOCHAM with the cooperation of Door Darshan prepared cassettes. It also looked for energy audit and for alternate source of energy like solar system and integrated energy policies. The energy audit would find potential energy saving areas. Another body of PHD Chambers of Commerce also are in the line and are implementing energy base programmes of the UNDP and EEC sponsored by the Energy Management Centre. The EMCAT project of US Aid is being looked even through IDBI.

The non conventional energy sources are helpful for the remote areas like hills where fuel wood could be saved. Such energies are wind energy, small hydro cogeneration, bio mass combustion and solar energy. Other forms of renewable energy are tidal powers, sea wave power, biogas and ocean thermal power. These energies can help in terms of environment protection, providing employment and social security. The estimated potentialof Renewable Energy Sources in India are as:

Biogas Plants	12 million
Bio mass	17000 MW
Improved Chullahs	120 million
Solar energy	5 x 10^{15} KWH/year
Small hydro	10,000 MW
Wind power	2,000 MW
Ocean thermal	50,000 MW
Sea wave power	20,000 MW
Ideal power	9000 MW

The solar energy centre is utilised in various sectors of domestic, power generation, industry and transportation. The solar disk can produce 50 KW of power of village level. The day light is one of the important criteria as source of energy and such energy is pollutionless and reduces use of fossil fuel, coal, gas and oil.

In the building industry more emphasis is being provided for the building to be energy oriented. The solar energy is available mostly towards South West side in the Indian context and so the buildings are made to face that side. The frequent power break downs even at this stage has forced the residents to arrange their own personal convertors or temporary arrangement systems for electricity. The day is not far off when solar antennas would be found at every roof top on account of its being cheaper and pollutionless.

The energy conservation is more concerned in industries at four stages of production, conversion, transportation and utilisation. A cost benefit analysis of this needs to be carried out. The power loss average in the power sector is 22% to 45%. To save energy utilisation of waste heat in industrial process, bio methanation of industrial effluents and utilisation of industrial waste bio-mass is desired.

In the agricultural sector for pumping, most of the energy is spent. At present 17% of total energy is in the agricultural sector of which 85% is consumed by pumps. About 120 lakhs pumping systems already exist and 10 lakhs are added every year and whereas 64 lakhs are operated by electric motors, the rest are run by diesel. For efficient energy the selection and installation of pumping system, operating conditions of pumps and prime motors should be good. Moreover, rectification of existing pumping sets should result less power consumption. The quality is to be maintained as per the ISI codes and along with this an awareness needs to be created.

The animal husbandry too plays an important role in the field. From the sun only about 1% of sunlight is absorbed by the plants and rest 99% goes waste. Without crop residues no animal husbandry is possible and those are the greater pollution sources. The live stock goes to farms, gives milk, goes to man and man works. This is the energy cycle. From fire wood it is cheaper to trap thermal energy and dung used as fuel will replace fertilisers one day.

The National Productivity Council (NPC) that plays a major role is working for energy auditing, suggesting norms for developing sectors, guiding dual pricing for energy, and creating energy efficient technologies.

The involvement of IDBI is useful for energy conservation.It has been observed that energy losses are because identification of potential saving energy activities are not done, role of energy managers like in US are not realised as companies do not employ them. The legislative measures are missing and there is no energy labelling on equipments as to what is the capacity of that to be used. The demand side management can save greater amount of KW in demand. As per the research of IDBI it is revealed that the developed countries have energy as 10-14000 units/yr/capita whereas in our country it is 250-270 units and in the case that becomes 1000 for us then ultimately all sources of energy would vanish.

The growth of power sector in 1950-51 was 1713 MW, whereas in 1994-95 81,164 MW and electricity consumption was 7 billion KWH to 351 billion KWH respectively for those periods. It is understood that by 2005 it would be 1,40,000 MW. During peak the shortage is due to transmission and distribution losses. IDBI has floated debentures also in the market to raise funds. The IDBI further is looking for EPEC Equipment Finance for Conservation; EAS Energy Audit Subsidy, EMCAT Energy Management Consultation and Training being run by US INFUSE Innovate Funding for Saving Energy.

Today Allen Badley firm has devised many systems and encouraged automation to reduce energy losses. The SAIL, HINDALCO and many leading sugar industries are having energy conservation specialists appointed in their firms for efficiency measures.

21

Handling a Natural Disaster in a Developing Country

From the very beginning of the human settlements the natural disasters have been playing havoc with the life and properties of the various settlements despite man's extraordinary achievements in the field of science and technology; not much progress is found in the nature's fury from that time till date though at places it seems to have aggravated more because of the greed, vested interests and corruption which has been much intensified.

Do you think in developed countries compared to developing ones any measures have been instituted wherein they could forecast such disasters except few cyclones or floods? Or else is it that if at all disaster has taken place then in no time relief measures had started?

As usual the things are common to all countries. There is a sort of formality which every nation does in the name of its system. If you watch seriously, the steps everywhere are found to be exactly the same. Navy or Army is called out as rescue measures or grants are provided from national funds or voluntary agencies come into action.

In the beginning of the settlements the floods caused much harm. Historically the old constructions have always sprung up all along the rivers. The river Nile of Egypt swept away all settlements and to date even the actions or the measures for disaster management are the same. The countries which are close to sea coasts are prone to the ravages of cyclones. Though the floods and cyclones can be predicted and warning signals are given but still don't you think that casualties and destructions have not become less?

Of course, earthquake todate remains unpredictable and this is the only disaster that takes heavy toll of lives and causes much destruction of properties. Though more countries are now engulfed under the earthquake belt but do you think anyone is bothering to look for seismic safety measures while designing and planning cities? We find regular multi-story buildings with no margin for horizontal shear stress being left.

The Tornadoes are other forms of nature that sweeps away human beings along with their belongings. The cloud burst are occurring in the hilly regions most often destroying man and properties. The volcanoes eruption in volcanic areas though is predictable but still casualties remain over there. The wild fire is effecting nature's ecological phenomenon by burning out trees, animals and taking toll of human lives. Finally is the worst disaster of draught from which there is no escape as almost the entire African countries are in the grip of the same.

One thing is realistic that whenever such disaster occurs, the people's sympathy is generated towards relief measures which is good as well as healthy and the funds are collected in cash and kind. However, misuse of funds starts simultaneously which is not healthy. The relief measures need more emphasis in terms of practicality because this is the sphere which has to be activated almost in all disasters.

Never is it seen in a film or any media where the citizen's role is directly or indirectly portraied. Perhaps don't you think presence of mind is the foremost aspect for those trapped in such disaster that can help in disaster handling? The media could better guide them. Otherwise, most of the poor people for whom earning two times a bread is the main moto of life, for them such things do not matter much.

The integration of various sections or Govt. Ministries is required so that at any time one authority could move for relief protection rather than so many bodies doing the job in their own style and not in coordination. Don't you think it is some what to be related to global level and better would be if a body could be formed at that level for quick decision on funding and supply of medicines.

Even if the prior information is provided about a disaster, people take that lightly till they are actually trapped. The warnings and precautions must be taken seriously and rescue measures should be on the move at once.

Even very rich people who are trapped can be seen behaving like beggars at those critical hours as at that time it is God's help who so every may be there. The author recollects one such hour of 1978 Yamuna floods when all including effluent people came up to the terraces of the roofs to collect what helicopters were dropping food stuff from the top.

Do you advocate that the integration of a disaster area to the town, city, country and international level is desired to be formed and a system to be developed. Within 5 minutes the town level services may operate. In 30 minutes the state level, in an hour or two the country level and within 4 hours international level help may reach the spot. All measures of finances and other sorts of help may be included in the services. For that rehearsals or mock exercises are required. This may be given priority and every block should have a provision of rescue measure and even while sanctioning plans such measures be cared for. The city's population need to be divided into various wards and districts in terms of disaster management.

It is desirable to monitor failure and successes of the past happenings. In Uttarkashi during the earthquake in 1991 there was so much rush that people were not allowed to go up beyond a certain point as that would have invited huge crowd which itself required its own management. The voluntary organisations and the rich entrepreneurs can come up openly to help the masses and such bodies can be a great help.

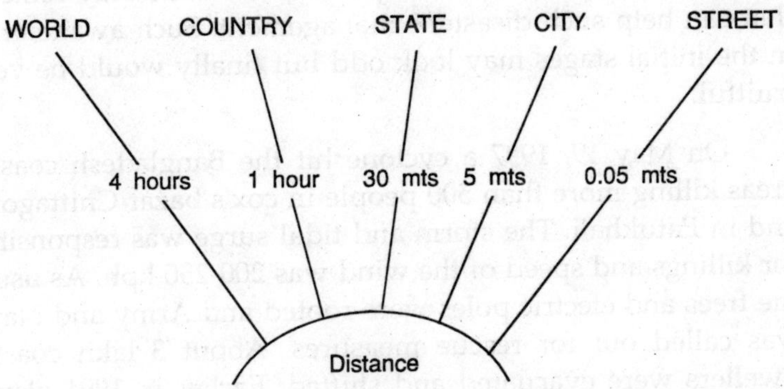

HELP CHART

Fig. 21.1

The establishments if made, may be voluntary, involving retired people or senior citizen wardens or unemployed youths who could be appointed on monthly basis. The aids shall be for food and medical requirements. Don't you think different strategies could be helpful depending on the type of disaster and the area affected. So let the idea of helping each other be developed and adequate training be imparted. After every 100 yards there may be a relief stall established to take care of such eventualities, or temporary relief sites be proposed in cities as planning measures for such operations. But there is every possibility of getting the same encroached. Those may be put under defence operations to keep them reserved. Even the financial expenditure spent on an area could be even repaid and if this becomes practical then even top world financiers would not be found wating in making all financial help available.

In a developing country context where the Apex courts take up the issues and in no time interfere, such missions are bound to be faster and to the point. Otherwise the authorities pay deaf ears and there are more pilfriges and instead of help nuisance is created in terms of theft and other negative practices. Says Mohan's concept that this is the only remedy that can help such disaster's management. Such awakenings in the initial stages may look odd but finally would be very fruitful.

On May 19, 1997 a cyclone hit the Bangladesh coastal areas killing more than 500 people in cox's bazar Chittagong and in Patukhali. The storm and tidal surge was responsible for killings and speed of the wind was 200-250 kph. As usual the trees and electric poles were rooted and Army and Navy was called out for rescue measures. About 3 lakh coastal dwellers were evacuated and shifted. Earlier in 1991 also a cyclone in Bangladesh killed 1,39000 people.

The Godawari Distt. in the Andhra Pradesh too is prone to cyclone and during one cyclone of 6-7 Nov. 1996 and the 7th cyclone, about 1000 people were killed and more were missing. It was in an area of 55 km with a wind speed of 220 kmph. There were rains of 26 cm height for 6 hours and the total waves were 3 m high. The damage to crops, buildings, power supply and communication was immense.

The cyclones mostly occur between April, May, October and December. In India the cyclone in Bay of Bengal originates near Andaman Sea or in South China Sea and moves through east coast of India between Tamil Nadu and West Bengal. In Arabian Sea it originates near Lakshdweep and moves in North East direction hitting Konkan and Saurashtra areas.

The effect of cyclones on the built up structures of industries and low rise buildings is of flying away of roofs and wall claddings. The failure of poles is very often above ground level and is due to substantial portion of that being on account of horizontal loads of unequal magnitude. The mitigation measure account to, use of J bolt to replance U bolts, steep slopes on roofs with least overhangs and good foundation system. The poles require concentrative load. The cyclonic shelter are required for every 100 families with good amenities.

The cyclones are predictable and strict orders are required to be framed parallel to curfews for the people to abide. The relief measures may start from the time of declaration of the warning. The camps and evacuations may start immediately. The aim should be to reach zero casualities.

The efforts may be made to keep restoration of electricity and water supply as usually it is observed that the two remain disrupted which creates more problems. Even fire also takes place in many places which requires control. After the warning signals people may be told what to do and what not to do.

During April-May 1997 two severe earthquakes were recorded one in Iran of intensity more than 7 on richer scale and one in India on May 22, 1997 in Jabalpur Distt. of MP of the intensity of six on richter scale lasting 30 seconds when three continuous tremours were felt between 0420 hrs. to 0430 hrs. lasting more than 30 seconds. The relief measures in the form of payments were to be made immediately to the families of deceased.

The other damages caused by the earthquakes of the world were in Bihar and Nepal in 1988, Uttarkashi 1991, Latur Usmanabad in 1993. The state of Bihar faced the worst earth quake in 1934 and 1988 of intensities 8.4 and 6.5 on richter scale. The studies revealed proper detailing of reinforcement to ensure ductile behaviour of columns was not done. Another study showed cracks due to corrosive nature which the earthquakes only aggravated. The Uttarkashi earthquake of 1991 was of magnitude 6.6 on richer scale in which more than 1000 people were killed and 3000 houses were demolished. The sloping roofs showed horizontal cracks due to horizontal twist caused during earthquake motion. Actually the remedial measure is that horizontal bracings are required. The masonary structures due to non-setting of the building on firm soil due to violations fell down or developed cracks. Where the foundation system could withstand as a single unit things have stood well.

The Latur earthquake of 1993 was 6.4 intensity on richter scale that tolled about 10,000 lives. The mud and stone walls collapsed. The stone masonary work which was weak in tension collapsed, and cracks developed thus reducing the effectiveness to resist horizontal sheer. The need of the time is to use one visco elastic dampers. The ductility and energy absorption capacity of structures is essential that reduces cracks.

Earthquake of Assam 1987 was of 8.7 intensity being of max. intensity combined with that of Garhwal and Latur. The recent one at Coba in Japan destroyed the economy of 2nd largest city of the world in seconds.

The earthquakes are caused generally on account of

1. alteration of earth's crust by corrosion and sedimentation (10" in 9000 years).
2. On account of releases of heat energy in the outer crust of the earth due to delay of radioactive materials. The decay leads to fusion generation of force, rise in horizontal and vertical movement and causing cracks.
3. The slow cooling of earth's surface leading to contractions.

4. By creation of large dams in areas like Las Vegas, Zambia, Rhodesia. The Greek faced many such tremors after dams construction.

5. Earthquake under sea is called marine earthquake. They cause damage to coastal areas. The highest tidal wave in 1737 was 213 feet.

Delhi falls in severe earthquake belt. More than 5.3 observed on richter scale are the earthquakes of July 15, 1720, Aug. 31, 1803; Oct. 10, 1956; Aug. 27, 1960; Aug. 15, 1966. About eight earthquakes of medium intensity 2.5 to 3.6 were recorded during 1992-96. The seismic belt is Delhi-Hardwar ridge between 28° to 30° North and 76° to 79° E that coincides with Aravalis beneath the allusival plains of Ganga basin of NE of Delhi towards Himalayan belt. At Rohtak it has an Aravalis fault zone. The major fault zone passes through Mathura-Moradabad.

The earthquake experiences have revealed that always avoid L or U shape planning and and better adopt a box shape. The dead lead of a building should be min. The centre of gravity to be as low as possible for stability measure. Avoid cantilevers, large openings, stone masonary work and vaulted roofs. The foundations should be better of on firm ground. The frame construction with horizontal bracings is a better design. The concrete ring beam may be built over ground floor and slab roofs. The building materials used should be light weight and fire resistant.

Immediately after earthquake it is found that there is danger of fire. In San Francisco's EQ of 1906, 95% people died due to fire.

On 26th January 2001 a severe earthquake of 6.8 mark rocked Bhuj State of Gujarat. It tolled about 20,000 lives and was severe. Again, in June 2001, an earthquake of 7.8 intensity rocked Peru city of South America.

Floods

The floods are a constant menace particularly in coastal areas. Now-a-days of course dams have been built for storage of water and retaining walls are built as bandhs in the cities but not much results are seen in the total scenario. The floods forecasting techniques are also quite accurate. But mostly one has to depend on the after flood measures as the areas like Brahmaputra basins and likewise other rivers where crores of people live on its banks alongwith animals every year there is a heavy casuality.

The floods are caused due to block of the of flow and overflow of banks. The floods occur in Delhi due to Yamuna and are most frequent. The danger level of Yamuna is 264-83 m which has been crossed 25 times during the last 33 years upto 1997. There has been high scale floods in 1977, 78, 88 and 95. The flood zoning covers an area of 199 m to 212 m. The flood central order divides NCTD into 4 groups, Sahadara, Wazirabad Babarpur, Alipur, Nangloi Najafgarh. The channelisation of the river Yamuna desires to reclaim 8000 hectares of land which is dangerous. The role of vegetation is maximum to control floods.

The Himalayas need maximum protection against hazards. The hill areas are prone to landslides, high speed winds, heavy rainfall, cloud burst, flash floods and most of the area falls under seismic zone. The region stretches from Kashmir and Himachal Pradesh to Arunachal Pradesh and Nagaland. The Tehri dam has remained controversial on the earthquake issue for which Mr. Sunder Lal Bahuguna the world Chipko leader had gone on fasts many times. The dam is believed to have started on safety intensity of 8.5 as on richter scale which is not much and if that burst can flood a large area of plains.

Landslides

The landslides are a common phenomenon in the hills

when mountains come down blocking the passage of traffic and general life for days till wastes are removed by army on high peaks or by the authorities. The Himalayas, North East Hills, Western Ghats, Nilgris Hills and ridges are prone to such effects. During Sept. 95 in Kulu distt. of Himachal Pradesh, the floods were considered as a prima facie in landslides. The primary geological surveys are essential for an area before development. That would be appropriate to construct lined catch water drains, road side drains and for protection. The vegetation promotion techniques are required.

In the Himalayan region the stone quarrying operations, the destruction of vegetation cover, the heavy rainfall, toe cutting of valley side slopes are causes of land sliding. The provision to provide cross and radial drains are suggested to minimise the damages. The Nilgris landslide of 1993 burried a mosque, bus passengers, and damaged many such things. In Konkan area railway failures occurred during construction of railways. The remedy is to provide and develop software and stability analysis of slopes technology and forecasting landslides timings.

Cloud Burst

The cloud burst is an occasional feature in the hills. The floods of Alaknanda 1970 and of Bhagirathi 1978 occurred due to cloud burst in the upper Himalayan region. The cloud burst is therefore not a city phenomenon. It arrives with the speed of a thunder and lasts for a few minutes. On 16 August 1991 the cloud burst took place in Chamoli district in Dewar, Khandra village at 1600 meter altitude. The cloud burst and landslides both occurred there. The lightening brings heavy rains. This is caused due to rock erosion and new soil cover. The lightening rods above 7500 meter height are advised to be put on hills as those are prone to cloud burst. During the cloud burst the life of the area gets paralysed. There is casuality of men and animals and the kaccha houses break down. The rain water gets collected which takes time to evaporate. This also needs prediction techniques.

Avalanches

The danger of Avalanches is mostly in the higher altitudes where glaciers come down causing damage to the life of people and animals. From the slides shown in one of the seminar it was revealed how to reduce the quantum of a thick ice-berg. The artificial trigging (use of explosives) was thought of. The ice mounds were created to disturb the shape of the ice. This also needs forecasting and requires intensive afforestation to check it. This will require intensive research and training.

Forest Fire

The forest fire is a common phenomenon. In UP recently there was a big forest fire that remained for months thereby producing smoke in almost all villages of UP that made life unbearable. Some said it was deliberate but God only knows the cause. The wild fire is also caused due to earthquakes, cyclones, volcanic erruptions, hurricanes, floods and landslides. Also these are caused due to high winds, combustible fuels, pine belt, by the poachers and on account of delay in monsoon. The direct and indirect measures to protect are required. The fire fighting and chemicals are required for such purposes. The mock exercises and identifying landmarks are required to solve the issue. The direction to evacuate the hot spot zone is a must. The point is in which direction people and animals should move at the time of crisis. The colder zones require indication. The wider paths may be marked in all directions straight and radial for escape. A fire proof watch tower be built near rangers house to watch the quantum of fire and give directions to the habitat whether those are men and animals of the forest. It is understood that the best is to move away from the scene as all training and researches be focussed on this issue. The traffic diversion, and other immediate actions need wise tackling of the problem. The wireless medias, cellular phones could be better techniques for contact outside fire zones and for assistance and guidance for evacuation. The important office materials files require fire proof cabinets and almirahs. This training, detection, the fire brigade and transportation needs require to be paid attention.

The Delhi Fire Services is the one that deals mostly with fire cases and fire arrangement of various buildings within capital. Its task is gygantic. There are 1986 personnel working there and 165 vehicles are at its disposal. In this field no NGO is active.

Tornados

The tornados are not so frequent but some times occur in the form of high speed winds that blows everything before it. An inverted funnel shape figure could be watched in the sky even from far off areas and this funnel is created by clouds and high winds.In late seventy people in Delhi watched one such tornado in Delhi University, Maurice Nagar area which vanished along Najafgarh drain and the river Yamuna. There were many casualities and persons on two wheelers were seen flying high in the sky andthen ultimately missing.

After Effects

The after effects of disasters are awfully bad. The Latur earthquake left the epidemic plague which spread all over the country and people were found wearing a mask everywhere. The breaking out of a fire after earthquake is a common phenomenon. After floods too care is taken that epidemics does not break out. Normally DDT is spread on roads and so many other measures are adopted that require proper examination. It is a vast subject in itself. So we have enough medical cares after such panic but any such revealing mechanism needs to be properly thought of.

The only way in a developing country to make such operations counting zero casualties says Mohan's concept, is the intervention of Supreme Court and tapping experts through them in evolving various systems. Otherwise if left to politicians or the authorities the results would be neglegible as no accountability is there and wastage and theft of funding very much prevalent.

Finances

The funding requires experts through courts. The revolving fund even at global level be thought of and let rich entrepreneurs come forward. The money is to be repayed with interest. And that is the only solution meeting such tragedies says Mohan's concept. Otherwise why would any one finance. At least at odd hours finance can solve many such problems. The global level finance should be supplemented by centre and state level and enough revolving funds collected. If in an election the leaders can accumulate a huge finance then why not for such missions? This canbe managed very welll if there is a will.

As per one of the experts in India the total funds available under the state and central share are Rs. 6304.24 crores and Rs. 700 crores under Natioal Relief Fund. More is required at local level and Panchayats.

An Action Plan needs to be formulated for disaster management and may include the following:

1. What is to required at Central Government or even at Global level.
2. The suffering of people need to be made less.
3. The time gap to be minimised.
4. The long term and short-term solution be thought of. For earthquakes and cyclones long-term plan is desired. For floods both short and long-term solution is desired and for landslide wild fire action plan at local level is desired.

5. Alert signals may remain always activated.
6. The training needs to be imparted.
7. The media should bring out alert and warning signals.

22

The Indian Tourism Industry

Ever since the dawn of civilisation right from the Vedic age and Aryan periods history reveals that Indian soil always remained as an attractive place for the foreigners because of several reasons; may be due to its unaccountable wealth, rich heritage; its vast density or its people; whatever was the ground that constantly persisted for years in line with China, Japan, Sri Lanka and other Asian Countries reaching its peak at one time before the downfall of tourism started. The tourism management suffered a severe set back throughout the world on account of economic recessions, due to epidemics most often breaking out and due to other causes. We may remember that during Asiad games 1981, on account of dangue fever, very few tourists visited India and in the World Cup Cricket, 1996 when bombs blasted in Sri Lanka; several world players refused to play in that country.

At a time when the concentration of tourism was evaluating the damages caused to the tourist industry; there was no simultaneous thinking on raising parallel level tourism spots. India lost one of its charming tourist state of Jammu and Kashmir due to terrorism as a result of which the income of various persons depending on tourism sustained heavy losses. However, the tourists started flocking to lower upper Himalayas and Himachal Pradesh in place of Kashmir. The European tourists are scared of the Indian summer climate which at most places is dry or semi-dry and in the winter months they find it more convenient to come over to India.

The Tourism Industry of other countries have involved the entire world. From any country a package could be arranged by paying commission to commission agents and those are organised so well that security, convenience, economy and confidence all remain uppermost. The Indian Tourism some how lacks that and so it requires a big propaganda, incentives and agents throughout the world.

Amsterdam in Holland is the most attractive tourism spot of Europe. One who has been Amsterdam is considered a tourist of international standing. There are tourist spots of wind mills, locked sea side, canals, the old Amsterdam town, the art galleries, the sex world, the diamond jewellery manufacturing and identical interests. Such world tourism interests require tabulation and then developed as per Indian scenario so that the identical interest of particularly European community could be identified and developed. This is because the maximum amount of money flow is from the tourists only, in dollars and pounds. Of course, from Arabian countries the tourists too are visiting India and that their needs should also be taken care of. The tourism is required to be healthy, meaningful and of economic interest to the country.

Agra, Delhi, Japan, Jaisalmar and other Rajasthan towns are having ancient monuments where tourists often go. In Delhi it is found that mostly the monuments are either crowded by car parking or shopping alleys. The tourists are not crazy to see our car parking. They spend money to enjoy in the real sense. It is found that they do not get fair deal from the shopkeepers. The same commodity they purchase for double and thrice the price which we get much cheaper. That pinches them. There are no standard eating and economical shops and also the lodging arrangements. At least this you do not find in the European context. They are very fair and tourist guides are there at places for helping them out. Very cheap food is available there. One of the very busy crowded bazar of Delhi, in India, called Paharganj, surprisingly has totally been changed as tourists bazar and all shops, residences have been converted to hotels for lodging and boarding.

Identically such surveys where natural tourism flourishes is to be watched carefully and the desired infrastructure could be provided so that the area comes up. There is no use of creating new areas requiring lot of finances but it is better to develop the existing areas where tourist flow is already there. The incentive and facilities could be given to the people living there who are willing to take part in the mission. At all India level, the grading of tourism may be seen in reference to a perspective plan of 25 years and that may even include areas of tribal living areas.

What is found in the Europe context that maximum paper work distribution and advertisements are done on the tourism which however in the Indian text is negligible. The sufficient attractive pamphlets of Indian tourism spots are desired to be made available with all tourist shops throughout the world free of cost which would accelerate the working of industry. The tourist movement is desired to be made well-known to all tourists and all needful infrastructure be provided. For example, if a tourist desires to have a look of Shahjahanabad then even tourists lanes could be decided where to move and even the guidelines can be provided on mode of conveyance. The expensive travelling is desired to be curtailed to economic fares with easy accessibility. In large hotels where such tourists stay, a fair dealing is desired and this is to be seen that in minimum litigation they may be involved. There can be passport problems, visa related exercises that require assistance to be provided from place to place. Even misplacing of passports be also helped out. The environmental tourism requires to be more encouraged as it would help in two ways; number one to protect the environment and secondly sometimes tourists are specifically crazy to go to the Safaris, the Zoos and the botanical gardens.

The health of tourists must be required to be seen on priority basis. Many tourists suffer from AIDS and other dreadful diseases. Of course, checking is not desirable as sometime back UK did the virginity test of women which was not appreciated. But even then dispensaries may be made available to them as many times they fall ill and cannot afford costly medicines. So the health issue must not be overlooked. The Tourism Ministry itself must have a health deptt. of its own rather than depending on Health Ministry to come for rescue. Beside the international tourists, the national i.e. inter-state tourists within country too are increasing and as such need due care. More hotels and motels at economical rates are desired for them. The circuit houses that accommodate government families are smaller in sizes and require expansion. The STD and other telemessage sources be made available to them. In case one desires to return to his country and cannot afford the agencies may come up to help needy persons, so that they may go back to their native town safely and happily. A country's image is reflected in serving such missions.

There can be rackets, mafia gangs, drug trafficking that require proper securities arrangements. That hampers the free flow of tourist industry. There are the strict laws of a country which are above all. The Tourism industry is required to bring up and particularly case for backward states so that their economy could be improved.

Concludingly the recommendations are as under:-

The scope for arranging pre-package tours from other countries be examined.

The free distribution of pamphlets and advertisements are required.

The security, convenience and economic issues are required to be examined from a tourist angle.

The scope for developing new tourist spots are required to be examined at national level.

A perspective plan for another 25 years of tourism is required to be chalked out.

The environmental tourism is required to be given top priority.

The specially backward areas need to be exploited for improving their economic conditions.

The sufficient guidance pertaining to visas passport may be provided to them at free of cost.

The rich entrepreneurs are required to come forward for helping out destitute tourists.

The terrorism, epidemics and other causes damaging tourism need to be improved. In case this happens the parallel arrangements be immediately made.

The pollution issues to be checked for healthy tourism so that tourists could remain healthy. In case they fall ill sufficient care to be provided and the Tourism Ministry should have its own Health Department for that.

The concentration of the tourists is desired to be at one place for security reasons.

There should be sufficient bazars and visiting sites to keep tourists busy.

For tourism there should be a very high class infrastructure of roads, electricity, potable water supply, sewage and other facilities. A good amount of budget may be kept for development of such facilities.

23

University Tours as Tourism Industry

Many eminent personalities who had held important portfolios in Government or else happened to be attached to any university either as a student, a teacher or else in some other capacity, or campus could not get that prestigious recognition as it deserved like Nalanda and Taxila of past India. For example at Delhi, the galaxy of colleges; St. Stephens, Hindu, Miranda, Law Faculty, Arts Faculty are exemples of the educational institutions of the country since prepartition era. The grand architecture of the British style buildings in the campus reflect a rich taste of the early builders.

Unlike other Universities like Cambridge and Oxford in UK; Delhi University campus remains unexploited. Though in Delhi, the university and colleges have been decentralised but still the Maurice Nagar area is a reflection of traditions of the age old traditions of friendships and rivalaries of the past era; for example the Hindu and the St. Stephens that always stood edge to edge in sports, education and other fields in the past, retain the same status till now.

The complex was beautifully chosen by early educationalists near the side of coronation pillar in north and the beautiful Aravali ridges in east alongwith Old Secretariat that stood on the banks of river Yamuna. Still one could see the Cavalary road where galloped the horses of British era.

The Najafgarh drain which flowed close by did not draw that much sullage and the entire area was lush green with birds and animals. However, with the openings of more colleges in the campus, the congestion started taking place.

A look of the restaurants where still subsidised items can be had still follows the traditions. The glittering young collegiates and motorbikes running at full speed; noise everywhere appears refreshing.

Visiting the campus area on week ends is a suggestion to tourist deptts., responsible for exploiting tourist spots in Delhi. All these colleges require to be protected like monuments. Like Tourist Guides in U.K.; they may be deployed here also. For the tourists stay in nearby areas of Kingsway Camp, Kamla Nagar and Timar Pur; cheap resorts may be hired or built.

The Aravali Ridge and the River Yamuna be focussed as main attention of visits as very few countries have that combination of so large spanned rivers and protected Ridges. The Ridge that side is full of history of Raj period, the Bounta, Ajit Garh, Ashok Lat, Hindu Rao Hospital, and Metcalf House and historical places over there. Of course, the Oxford Campus had religious churches attached to each college which here would be missing but other environmental aspects would compensate for that.

During flower shows during February; the convocations and on special days, the area is bound to get intensive tourist population not only from All India basis but from every corner of the world.

This would improve the environs of North Delhi which is very deficient in terms of recreational spaces and at least those residents of Delhi would also come to this spot.

With the tourist boom, the adjoining areas and economic conditions would definitely improve and Delhi Darbar which was proposed earlier in 1911 near Village Burari, once more organically will show that Delhi's siting at this place was really justifiable.

The University and College Authorities as well as aluminies would too like the proposal and make it all the more successful, with all kinds of support, finances and help. This would provide a strong economic base to the University Authorities which is grumbling since long for not operating so many needy issues on account of poor finances.

The Tourist Deptt. of Govt. would not have to invest on any major infrastructure in this area. In that case, there is a need to link river Yamuna to Aravali ridge near Khyber Pass side by an underground passage or flyover on adjoining Sham Nath Marg which is author's concept and is being favoured by Green Party of India; by directing this through a show of human chain. The Najafgarh drain which is at the back of Khalsa College too would start improving.

The side effects of the proposals of course can not be overlooked as more congestion of traffic and public is bound to be there as roads and infrastructure initially is not on that scale but gradually as income would start being generated things would be simplified. It is expected that 10000 persons would visit the river, ridge in the area during holidays and about 2000 persons in morning, evening and on other days of the week.

Viewing all that this would serve multifarious functions and better ecological conservation measures would be in sight. This would be an international tourist resort by 2025 A.D. for which even scope for a small airport proposal nearby may also be examined in north direction.

A one voice of public, authorities and politicians be also taken on this issue and let parliament debate on this and evolve a healthy management for this project of national value which is indirectly regarding conservation of ecological measures for the city. The NGOs could meet and decide its future prospects by working in coordination with tourism authorities.

The activity is desired only on week ends so that the public could relax in a western style. Rest of the days, this is not possible as university rush is there.

24

The Project Management

Case study - Narela Sub city, Delhi, India

This is one of the great need of the time to make a needful project meaningful and workable for the society. For this the finance has a major role as how to complete a project and then the role of bearing establishment charges and benefits, the later occuring after years. Most of the projects like Electricity Boards were earlier going under heavy losses and even to repay the loan was difficult. The World Bank too is charging interest for the loan it subscribes.

It may take as much time to conceive a project with all minor details that makes execution very quick. A fine report writing plays a major role as it has to cover all points. The reports on various subjects is different but contents almost remain the same.

For the town planning and architectural jobs the calculation of population is the main issue over which depends the distribution of commercial and industrial activities. The traffic generation, width of roads and its linkage in the settlement pattern, the land acquisition analysis and payments to be made are to be covered up. Nothing is visible even after spending crores on infrastructure electricity cables, water supply lines and sanitation.

On several occasions the Court cases come up when the entire project gets held up. Sometime Court cases are for rescue and protection of the interest of the Authorities and even gives incentives for early planning and execution. For planners such attitude to be adopted is difficult as they don't have powers. Therefore, at times Apex Court decisions are welcomed but sometimes those appear to be heard for bureaucrats.

There are required to be viability tests of various issues to know that whether a project is practical or not. If it is not practical then it needs to be dropped because there is no use of wasting money. Previously such notions were ignored but now the Authorities are very conscious about this.

For example a report is enclosed of the project Narela at Delhi covering 7365 hactare with NCR boundary. After approval of the local committee and authorities it goes to public for NOC and then finally gets approved through the concerned Ministry. The report is attached as:-

Draft Zonal Development Plan for Narela Subcity

According to the Master Plan of Delhi - 2001 Delhi is divided into 15 zones out of which 8 zones (A to H) are in urban Delhi, 6 zones (H to N & P) are in urban extension/ rural areas and Zone-O i.e. Yamuna River.

A Zonal Development Plan is a plan for a Zone. The Zonal Plan (Divisional Plan) details out the policy of the overall Master Plan and acts as a link between the lay-out plan and the Master Plan.

MPD-2001 stipulates that already approved sub-zonal plans in confirmity with master plan shall continue to be operative. The proposed zonal plans after approval should supersede earlier approved sub-zonal plans.

Section 8 of Delhi Development Act provides for preparation of Zonal Development Plan simultaneously with the preparation of Master Plan or as soon as may be, the Authority shall proceed with the preparation of Zonal

Development Plan for each of the zone into which Delhi may be divided. However, a Zonal Development Plan may contain a site plan and land use plan with approximate location and extent of land uses both as public and semi-public buildings/ work centres/utilities, heads, housing recreation, industries, business, markets, schools, hospitals, open spaces etc. It may also specify standards of population density and various components of development of the Zone.

The procedure to be followed for the preparation and approval of Zonal Development Plan is laid down in Section 10 of the Act.

The Zonal Plan of the article has been formulated under Section '8' and will be processed under Section '10' of the D.D. Act 1957.

Narela is located on the West of G.T. Karnal Road and spreads over in 3 planning divisions i.e. part of zone-M, part of Zone N and part of Zone P.

Following are the boundaries:

North	:	NCID Boundary
South	:	Boundaries of Sub Zone C-21 and H-19
East	:	G.T. Karnal Road and its bypass (NH-1)
West	:	Western Yamuna Canal

The total area of the Narela Subcity is 9866 Ha with the following break up:

Area under part Zone = M	:	2257 Ha
Area under part Zone = N	:	2767 Ha
Area under part Zone = P	:	4842 Ha

Population and Employment

MPD-2001 says the urban extension areas are proposed to be planned with an overall gross density ranging from 160 to 200 persons per ha. and based on the area under consideration the proposed population computes to 1.4 million.

The MPD-62 has recommended to develop Narela as one of the ring towns around Delhi Metropolitan Area (DMA) owing to its location and development. It proposed a township for 60,000 people with government offices and industry to make the town a self contained unit. The proposal included the adjoining developments of Mamurpur village and the development undertaken by the Ministry of Rehabilitation. Broadly the landuse plan for 1753 acres was detailed out with a gross residential density of 50-60 persons/acre. However, it also mentioned that "the proposals with respect to Narela are only indicative and not mandatory."

The following are the provisions as per MPD-2001:

"To accommodate the balance 3-4 million, population the DUA-81 which could systematically hold 82 lakh population, approximately need to be extended by about 18,000 to 24,000 hectares over the next two decades to effectively respond to the growth of the capital. Land required for various developments in the extended time frame by the year 2001 may be acquired from time to time, with due regard to the balance development of the city. In the Plan, 4,000 hectares (approx.) have been added to DUA-81 urban limits, thus the balance requirement would be of the order of about 14,000 to 20,000 ha. During the preparation of the plans of Urban Extension Najafgarh, Narela and other important settlements in rural use zone like Nangloi, Bawana and Alipur shall be provided with infrastructure and integrated in the Urban Extension Plan."

The land in the Urban Extension (UE) would cater under different landuses in the following manner:

Table 1

Proposed Landuse in Urban Extensions

S.No.	Land Use	%age of land
1.	Residential	45-55
2.	Commercial	3-4
3.	Industrial	6-7
4.	Recreational	15-20
5.	Public and Semi-Public Facilities	8-10
6.	Circulation	10-12

Further few specific proposal have been outlined by the MPD-2001 to be accommodated in Urban Extension in North which are as follows:

(i) Integrated Freight Complex

"For the efficiently in goods movement by road and rail, freight complexes have been recommended which would consist of whoesale markets, warehousing, road (truck) and rail transport terminals so as to curtail the movement of heavy vehicles. Such complexes are to be located in the places where they intercept the maximum possible regional goods traffic entering the city. The four sites for freight complexes are:

At Madanpur Khadar (NH-2)

At Patparganj (NH-24)

At G.T. Road (NH-1)

At Bharthal (NH-8)"

(ii) Storage of Oil and LPG Gas

For storage of oil and LPG gas in addition to two sites near Tikri Kalan and Bijwasan, MPD-2001 states that, Third Major oil terminal site to be developed near Holambi Kalan on Delhi Ambala Railway line in the beginning of the next decade i.e. in 1991. This site could be connected to the existing oil pipe line through an 8 Km. pipe line. This storage depot shall be mainly for white oil and LPG.

(iii) Inter-City Passenger Movement

To cater for the inter city passenger movement in Delhi which has been growing at about 4% per annum, four metropolitan passenger terminals are proposed which would also serve the DMA. These are trans-Yamuna area, Okhla, Bharthal and in North Delhi to cater to the proposed Urban Extension.

Urban Extension Plan Provisions

The DDA approved the urban extension plan on 30.6.1987 vide resolution No. 79 which proposed the phasing of the plan based on the population projects given below:

Table 2
Population Projections

Population projection by	Population Urban 2001 (million)	Population accommodation in DUA-81 (million)	Balance (million)	Land need (hectares)
NCR	11.0	8.2	8.2	18,000
DDA	12.2	8.2	4.0	24,000
As per existing trend	14.4	8.2	6.2	35,000

Source: Extensive Modifications to MPD-2001, Agenda dated 30.6.1987, DDA

"The urban extension could be restricted upto phase II if the population of Delhi could be restricted to 11 million by 2001. In case of 12.2 million population by 2001, the urban extension would be upto Phase III and in case 14.4 million population by 2001, the urban extension would be upto Phase IV."

As already mentioned, Narela Subcity forms a part of Phase IIB and III of this Urban Extension Plan.

NCR Plan Provisions

The statutory board for NCR which was set up in 1985 prepared the first statutory Regional Plan 2001 which is in operation at present. The NCR region extends over an area of 34,242 square kms. in the four states of Haryana, Rajasthan, Uttar Pradesh and National Capital Territory of Delhi.

The relevant recommendations are as under:

The NCR plan has identified 9 major corridors in transport network system with the prime objecive of controlling growth of Delhi. Narela Subcity area falls on one of these corridors i.e. Sonipat - Panipat corridor.

The plan has suggested provision of 100 mtr Green belt along the National Highways and expressways and 60 mtr along the state highways. Further large area under Green belt has been reserved which extends from proposed 100 mtr expressway in the Urban Extension Plan - 2001 upto the NCT of Delhi to act as buffer zone between the urbanisable areas of Urban Extension proposed in Delhi and DMA town to Kondli, Haryana.

The Existing State

As per the provisions of MPD-62, the Zonal Plan of Narela area stands approved by the government and is known as Zone II. The salient features of the Zonal Plan of Zone II are as under:

The Zonal Plan in its proposals stressed on the need for re-development of the existing Mandi apart from providing a community centre, police station, a hospital and a college site. The proposal incorporated a green agricultural belt in order to limit the growth of township from urban growth and haphazard development during the plan period.

The township was designed for 1.05 lakhs at an overall town density of 55 persons per acre. The town has been divided into two parts i.e. Phase I and Phase II. The gross density varying from 100-150 persons per acre has been provided in zones in relation to the landuse i.e. a density of 150 has been proposed near work centres and density of 100 towards the outskirts of the town. In all a total of 841 acres was proposed for residential development. The industrial area of 534 acre proposed to generate in all employment for 16,000 industrial workers.

Physical Characteristics

The location of the proposed subcity is on a terrain which is ideally suited for urban development. The entire subcity is very well connected with the main city as well as the adjoining state of Haryana. The Railway line which is running almost through the middle of the area divides the subcity into two major zones.

The entire subcity area is expanse of more or less flat terrain with land gradually sloping inwards to the south. The area has a very large tree coverage in the form of orchards, forest (declared as reserved forest) and farmlands. The subcity area has good soil for construction. The whole of subcity has good natural drainage system and the area is divided into two drainage basins (i) catchment of Bawana estate and (ii) catchment of Drain No. 6. Both these drains have number of subsidiaries spread over the entire subcity holding mainly the rural discharge.

Existing developments

The significant existing developments in the subcity are as under:

(i) Abadis of 16 villages and census town of Alipur and Narela Urban Area fall in the subcity. A list of villages falling in Narela subcity with their population as per census of India 1991, is annexed as Annexure 1.

(ii) Large number of farm houses are existing in this area. Number of unauthorised colonies/construction has also grown organically among the major rural roads within the subcity area.

(iii) Major chunk of planned development recently taken up by DDA is near the Narela Urban Area where roughly 5.16 ha. of land has been acquired and is in the process of development under various uses for providing facilities to about one lakh population.

(iv) DSIDC has acquired and developed 247 ha of land which adjoins this development land for industrial use.

(v) Land for utilities like water treatment plant, roads, ESS, part integrated freight complex has been acquired and is being developed.

(vi) There are a number of godowns and industries. Some of these industries do not conform to MPD-2001 provisions and have been ordered to be closed down by the Supreme Court of India (List of pollutant industries is at Annexure II).

FrameWork for the Preparation of Zonal Development Plan

In addition to the provision given in the Delhi Development Act and MPD-2001 with regard to the preparation of the zonal development plans, other broad objectives of the zonal development plan are as under:

(i) To incorporate the Zonal Development Plan for Zone II prepared as per MPD-2.

(ii) It broadly accommodates the land uses as envisaged in the Urban Extensive Plan 2001.

(iii) It retains the major calculation system as envisaged in the Urban Extension Plan.

(iv) To retain the agricultural green belt in between the

proposed urban development and the National Capital Terrtitory of Delhi.

Landuse

The landuse plan is based within the framework of urban extension plan, its existing and proposed circulation system and the different residential density pattern to be followed for the various residential pockets. The MRTS corridor identified along the existing rail corridor conducting rest of the city with Narela upto Holambi Kalan is envisaged to be a high density zone. Following is the land use break-up at subcity level:

Total area	:	9866 Ha
Area under agricultural green belt	:	2501 Ha
Proposed urbanisation area	:	7365 Ha

Table 4
Landuse Break-up for Urbanisable Area, Narela subcity

Landuse	Area (Ha)	(% age)
Residential	3,165	42.99
Commercial	566	7.7
Public & Semi-Public	875	11.9
Manufacturing	308	4.1
Utility	324	4.5
Government	16	0.2
Recreational	1,029	14.0
Transportation	1,079	14.7
Total	7,365	100.0

Residential Development and Housing Areas

The subcity when fully developed will provide housing

opportunities to over 2.8 lakh families of all income groups. It is proposed to have differential residential densities with rail line identified as the major high density corridor with next residential density of 600 pp ha. The pockets near the 30 mtr wide road and 60 mtr wide road will have net residential densities of 400 pp ha and 300 pp ha respectively. The development of residential areas will provide for housing as both public and private sector. Special emphasis will be given in reserving the area for the resettlement of squatters. The types of housing in Narela could be:

- Cooperating Housing Societies.
- DDA housing for its registrants in the form of built up flats, expandable housing.
- Janta Housing in term of one room tenements.
- Institutional housing for employees of various government organisations.
- Resettlement of squatters

The village Abadi areas, unauthorised colonies, would be dealt with as per statutory provisions and policy decisions taken from time to time while working out detailed plans.

Commercial

The sites for the district centres and community centres as per the MPD-2001 have been shown in the land use plan. Community centres to be planned in future will also have a component of service centre suitably identified at the time of detailed planning to take care of such permissable activities at sector level. Similarly the sites for Oil Terminal, Integrated Freight Complex have been shown in the land use plan.

Public and Semi-Public Facilities

Public and Semi-Public facilities upto the divisional level

(Zonal Plan Level) have been identified and located in 18 pockets in the subcity plan as per the norms given in MPD-2001. Each public and semi-public facility pocket is a cluster of various master plan facilities worked out as per the population of the sub-zonal pockets.

Since the subcity is divided in 2 parts by a major railway line attempt has been made to locate the facilities as per the requirement in both parts of the proposed development.

Manufacturing

The existing pollution industries as well as those which are not conforming to the provision of the Master Plan would be dealt with according to the decision taken from time to time by the concerned agencies.

In the new industrial are as the industries as per provisions of MPD-2001 will be allowed to be set up and part of the industrial areas would be used for re-locating the non-conforming industries from the city of Delhi.

Utilities/Physical Infrastructure

Component	Remarks
Water Supply	The total water requirement of the subcity works out to be approximately 120 MGD @ 80 gallons per day. Water Treatment Plants with a capacity of 300 MGD has been planned which would serve as well. The water supply system will consist of underground reservoirs command tanks for 1.5 to 2 lac population) inter connected by water mains. Each Command Tank will have its

	own command area. About 8 command tanks have been located. Land for MRTP Western Yamuna Canal has been acquired and is in the process of development by MCD.
Sewerage System	Central Sewerage expected to be generated is 90 MGD. The STP's have been planend STP-1 with a capacity of 20 and STP-II with a capacity of 70. System to consist of Gravity main, sewage pumping station and rising mains. The sewerage treatment plan towards the east of GT Karnal Road are in the process of development by MCD and would take care of requirement of Narela subcity.
Storm Water Drainage	Total discharge is of about 7,000 cusecs. The total area falls under the catchment of Bawana escape and Drain No. 6. Also there are number of subsidiary drains such as Bakner Drain, Sanoth Drain, Gunha Drain etc., which will form major networok for carrying the storm water.
Electric Supply	Total requirement : 1,4000 M.W. System may consist of 400 KV ESS (1 No. under construction)

	and 220 KV ESS (5 nos. 1 already existing) and 66 KV ESS (20 nos.). The detailed locations and numbers would be decided in consultation with DVB at the time of preparation of detail plan of the subcity.
Refuse Disposal	Total Refuse Collections is 840 tons/day. The process of disposal to be through Sanitary landfills composting and incineration depending upon the quality of the refuse. In future emphasis would be laid on solid waste disposal by more scientific method, where the land requirement gets reduced to minimum in consultation with the connected agencies.

Circulation

The circulation system of the subcity is based on mainly surface transport i.e. rail and road. The strengthening of the land network supported by MRTS would provide major public transport to the subcity from another city. The details are given below:

(i) Proposed Railway System

The rail corridor Broad Gauge Railway Line to Amritsar running through the subcity provides for goods and passenger movements. The existing railway corridor has already been electrified for the phased movement of the commuters as well as the goods from the nearby areas. As per the provisions of MPD-2001, in consultation with railways it is proposed to expand the main corridor to 4 tracks with a passenger terminal near Holambi Kalan. It is also proposed to extend MRTS upto Holambi Kalan terminal in the first phase of development which would be extended to Narela subsequently. Land for this purpose has been kept reserved at the zonal development plan.

The North West corridor is of about 20 Kms.

(ii) Road Network

The major network of 100 mt., 80 mt., and 60 mt. and 40 mt. roads as envisaged in MPD-2001 (Urban Extension plan) has been retained. The road system would be further connected by hierarchy of roads given as under as and when they are finalised:

(a) Streets : 28 mt and 20 mt without Green Buffer

(b) Streets : 20 mt and 12 mt.

It is envisaged that the above hierarchy would successively segregate the different traffic flow and roads with green buffer would also combat the ground non-pollution in the area. Attempt would also be made to provide dedication for the movement within subcity.

(iii) Bus Depot

Bus Depots are also planned as per the population details.

In brief those are the issue taken up for preparation of the report. The annexures and other details of approvals of land taken seperately are mentioned to update the report. The heavy industries requires to be shunted out is also discussed under .... The formats and briefs given is a frâme work for report preparation is essential before starting a project.

25

Rural Structure in Uttaranchal Region: Need for Sustainable Development

For a settlement there are only two words that could be mention. Either that could be rural or urban. In between the two there is a wide gap between the living, thinking, attitude and behaviour of the people. No one could be credited as good. Both have merits and demerits. But, it is found that when one comes to city from a village his chances to go back remain bleak; may be that he does know that is going to suffer so many hardships.

Why he likes an urban area is the craze to be called as superior villagers. The villagers too look at an urbanite with highest regarding as if urban areas are heavens. One tries to retain that false still end and cuts connections with village people, forgets village life and ultimately become a pure urbanite. The story is same with villages in the Uttrakhand Garhwal region.

The city migrants often get double minded; whether to continue like amidst cities with better living standard but with pollution environment or to go back to their villages and opt for a hard life rural life, of course with natural surroundings. Though, often former tempts them but still in their minds remain the charming in scape which they never forget. These villages are the settlements like dots spread over a big canvas.

In the Eastern Himalayas of Garhwal region the village settlements dedicated to single caste. For example the Nauti village is for Nautial's, the Thapi village is for the Thapliyals, the Dimri village for the Dimaries and so on. Then there comes a category where is caste families live. In the village Gaind of Patti Tala Dhangu, Garhwal there are Gwari Brahmin families dominating besides others of Bisht Khatris and the scheduled castes.

The ancestor houses are found at most of the places. In the past building materials were free of cost as there were ample wood, and mud in the natural so house construction was much easier. The little bit spending was on the roof tiles. The labour too was in the village Gaind many houses are found locked whereas several in the abandoned state. Perhaps their owners have migrated to go for good. The entire site is hilly and a motorable road gives to each individual house.

Now of course an Intermediate college is running over there which 80 years back happened to be a small school up to class one or two; so from it, we could judge the progress of the village that how slow it is. There is still no lighting arrangement and no proper drinking water system.

There happened to be a spring water source called Katlipani which lies abandoned now. The only source left, Naulipani, is insufficient. However, now they depend on tanker's water which is brought from Kotdwar, about 12 km from there.

The village is approachable from Rishikesh as well as from Kotdwar. From Rishikesh one can reach through a motorable road. From Kotdwar side however, there is regular motorable road to the village.

Along the bus routes at Matkhir many shops have come

up. About Rs. 30 a month rent is charged per room over there. Now the residents depend more on the shops for their daily food rather than on the cultivable land. The money order economy is the major source of their finance. The money orders being sent by employed youths in cities to their parents.

There are some tiny settlements that comprise of two or three houses only like at the Soar, the Sindi, the Dundyal the Patalgarh and the Khet where families live in total isolation.

The Garhwal is a land of beautiful landscape with red rocks. The greenery near the villages Silogi and Cholsen is marvellous where orchards have been developed. The Silogi school happened to be the oldest in the region where parents used to carry their children on shoulders thereby covering as much as 20 km distance on hills. It was a boarding school earlier.

The houses are mostly double storeyed. The upper one is called manjul and the lower one ubber. Mostly a small stair case is located in the centre to go up and down. Somewhere, however, three-storeyed houses too are seen.

As they keep cattle; so to save them from 'bagh' or panthers they keep them inside special rooms called sunny which is little away from the habitable rooms.

The kitchen is just like a room where the service is for a very short duration as they mostly remain outdoor in the field. It takes them hardly an hour to prepare meals. They burn cowdung or wood and so the kitchen gets full with smoke and they don't mind talking and eating in smoke.

Every room has a loft called dhepar or mazannine built of wood where they store all their tools and others household utensils and goods. The windows are practically nil. Mostly small peep holes are given. The doors and windows are well decorative and are mostly of good quality wood.

The tibari is a common varandah or 'baithak' adjoining dundyal or common verandah where sit most of the people of the community over 'hukka' and discuss the joint community issues. It is the most important part and is mostly a pillered hall closed by doorways. Then there is chotra, which is a projection at upper level where all the family members sit and gossip. The choklya is an adjoining open space to the houses. In the farms, farm out houses are built.

Before even building a home one used to think about its division that is its distribution among the would be heirs and which normally decided the room sizes as the elder child got more spacious from those of the younger ones.

For the ablution purposes still people go in the open land and thereby create insanitation. The Katli pani which Gwari, family used appeared to be disease prone as most of the family members using it got eye diseases or even blindness. Only one or two vaids are there in the name of qualified doctor. As such the villagers fate depends totally on the God whenever they fall ill.

The belongings of these persons are hardly any, except their farmlands, old trunks, old clothes little bit jewellery, farming tools and cattle. The cattle too, few family can afford as they have become expensive. Some keep goats and sheep though cows and buffaloes are kept by many.

A Post Office is located there and the Post Master is the village head himself. The system is not regular and many times corruption cases are detected.

Now banks are liberally giving loans to people for building houses, for growing plantations, for purchase of cattle but still not much progress is there.

The walls have niches for keeping little things. There are cub-boards and big storage drums for storing food and crops.

The stone construction is used all over, from inside of which they do 'lipai' with cow-dung plaster.

The children enjoy natural scenary and sports. They go to orchards and pluck fruits, fetch water and all that. They go for fishing in ponds, tanks and canals; go for hunting in the forests, do bird watching and all the natural games they play. Mostly the trees are found tied with hanging ropes where they enjoy swings. Even at places they go for grazing of cattle in deep forests. They are usually of sharp brain and have some relation or the other with the every household in the village. They are fond of story listening and often compel elders to tell them something about the new world. For army and police services they have special interest.

There appears to be a sort of a poetry all around. The women could be seen as usual over 'musal' and 'danda' or on Atta chakki or busy with such things. They are very friendly among each other.

On special eyes the whole village combine. All looks gay and colourful. There are invidual functions too like celebration of marriages or when they worship their devta, that is God, when they wake up whole night and afterwards sacrifice some animal. At those times the entire village become one. Mostly the floor is used for sitting and sleeping purposes. The walls at few places have carvings of god, rivers and ridges. The people are sturdy and active. Most of the people lead long life as it is pollution less environment.

The village in fact requires intensive care and for emergency at least a dispensary with operation room, qualified doctors, compounders and the chemist shop is required. Every persons' health is a state problem, which must meet priority. Secondly, school facilities in all villages be given, irrespective of the children attending the schools. The adults education must start. Thirdly, all infrastructures, like electricity, drinking water, roads, etc. may be provided with arrangements of Sulabh Shochalayas.

A guidance bureau be opened which may guide them for getting jobs or for securing bank loans, for sorting out disputes and for other such important issues. A TV set and radioset be provided. It could run on generator arrangement if no electricity is there.

The useful cassettes for upgrading their life be shown to them. A community hall may be built where ladies may be given training for sewing, knitting etc. The schools should be agro-oriented so that agriculture may dominate. Other vocational training too could be imparted by the media or the polytechnics.

26

Controlling Sprawling Rural Urban Villages in Delhi

Catching hold of the problem of rural urban villages in Delhi from head would just like be holding a mad horse by neck. As a practice one can visualise the havoc created by particularly urban villages in making a mess in the cities more particularly so of the planned development envisaged on paper and so would contiue the damaged pictures on canvasses of all present and would be metropolitan cities of India.

Asking the original residents of the villages, that have already sprawled and where the resident landlords had sufficient agricltural land sold to petty owners for building shops, residences or industries; a simple question; that did the villagers really enjoy those haphazard constructions and could they control that in case they desired so? The answer from the major quarter would be that since their land was in demand and as they expected heavy monetary gain; the opportunity was utilised by them. The middleman came into their rescue and made them rich overnight.

The other section of the villagers would reply that the compensation since was less and they feared their land would be acquired so they sold that in the open market.

Another possible reply from educated class of villagers could be that since they had no future plans for their village in terms of ecology or for improving socio-economic status of the residents; the later had no option except to sell the land; of course by which they even then could not improve socio-economically as the money they earned simply was spent.

Examining the three versions mentioned above of a simple question we reach to a conclusion that there can not be any compromise with the villagers on this issue since this involves money and Government does not possess that much surplus. Even if compensation rates of land acquired were increased in Delhi and one member of the family was promised to be provided employment; the situation did not improve.

Now coming to the point from above paragraphs that it is desired very much to educate the residents about the present extension of villages particularly in eastern Delhi at Mauzpur, Shakarpur, Patparganj, Khajuri and inform them about the negative aspects of their selling of land to petty owners who could not manage living in a human way and how they live under lack of facilities with menacing health conditions of shallow dug up hand pumps, without shopping, school and health centres. And another picture of possible health planned growth, could be shown to them as a contrast. That could be telecast in short films.

It is an earnest desire to tell them that the children who are suffering, are the wealth of the nation who would take the country's responsibilities tomorrow as why deprive them of a healthy environment alongwith ladies and old people.

The NGO and public participation could achieve little perfection, provided the Government machinery could mate with them. All the three combined with confidence can turn these villages as future environmental rescue zones when cities would rise to greater heights.

The advertisements by TVs, Press, distribution of pamphlets, may have to be given top priority in terms of budget so that meaningful results could be obtained.

The Mini Master Plan of villages on the maps could be a good exercise but more than that this should be related with the public as to what they want and to the budget should cater for those needful projects of the plan.

Concludingly the present chaos created by urban villages in terms of its expansions under *lal dora* or extensions have created unhealthy living conditions in cities that should not be repeated elsewhere. A need of the time has come when villagers, office Authority people, NGOs all are required to sit under one umbrella to evolve a befitting solution for them. The advertisement media for healthy living be liberally made and supported.

The villagers would need money; NGOs their service charges and authorities a planned development; environmental, that's all, the trios, if they try, can achieve the goal.

27

Can Ever Our Rural Areas Improve?

Can ever the living condition of rural areas be better, it is doubtful. Unless urbanisation drags, the rural areas for centuries will remain as they are. This will be due to poverty andshear non educatd people living there who fail in every mission just for jealousy, hatred or due to some dada of the village running illegal businesses.

Only drawbacks are noted when any development work is suggested for a rural area. The Government official's attitude is also not favourable and the politicians too have hardly any time to bother for those non interesting jobs. More over being far away from media no one listens to them as well.

Whosoever has tried to help had to were go back, as villagers themselves not in favour. And when villagers get biased they can go to any extent even to heinous crimes without bothering for the results.

Rather it is not advisable too to stick to one's mission despite there being so many repercussions. It is better to withdraw.

Then how can those villages improve? In the developing

country the situation is different to that of the developed areas. The regional plan needs good finance, total literacy and poverty alliviation programmes. Tell me why any doctor would like to settle in any village because if there is a casualty and people start blaming the doctor then the later even cannot stay for minutes there dispite his being a Government job. Though, Government had made compulsory also for bankers and such institutions to be made available there but one is scared and often tries to get him self transferred. On one occasion a banker was posted in tribal area. It was revealed that the trible's head dictated the terms of payments at the point of knife. Such incidents are not put up by the media but that is the reality. So any one can become a scape goat of the wrath of the villagers.

In terms of security no one can guarantee so why such personnel are being sent to such areas as a punishment? And who would bother for the development of the village then?

An institution is improved when we really volunteer and take care. They can be only persons from those areas. Only training needs to be given and let some one take care of the situation.

The weak law and order can not take care of the innocent people who plan to visit the villages. Then one may think that why one is taking so much pains. Other's will think he has some interest and that's why he is doing so. But in those circumstances the development would ultimately suffer.

The corruption at the end of Government offices hampers the villagers tempo for taking up any new work or job. Without their help even a thing would not move. That is the cause of check on all development works.

28

Improving Environment of Rural Areas by Providing Necessary Facilities

A deep void, observed in the rural areas these days, is in terms of the various essential facilities, which are either missing or else even if they do exist, they remain mostly unattended. The latter is more of a problem as in that case neither one can claim again for fresh items, nor the authorities get easily convinced regarding replacing them and allotting new one.

No doubt, there had been a considerable progress on the overall global rural canvass, in the field of education, as a result of which the basic education has been introduced at the grass-root level, by the media of the primary schools and its impact has been felt by our society.

Perhaps this alertness could remain confined to education only, as nothing beyond it could be looked up regarding provision of other essential services like health, family planning, markets, community halls, veterinary centres, the community latrines etc. whose needs are badly felt over there.

It is not that these programmes, cannot be worked out or else it would require heavy finances, for which the authorities would be reluctant. It could be solved in a moment, if the authorities and the public both so desire. What is happening is that, the public wants too much from the authorities, whereas the latter tries to remain at a low profile, as a result of which both never compromise and so scarcity in terms of every item continues.

Much would depend on the basic planning concept which could be programmed on a computer based on factors like population, area of the settlement, its nearness to the city, climatic conditions, social and economic status of the residents of the area etc. and simultaneously *vis-a-vis* fix the type and size of the facility to be provided for such areas. In fact, this would ultimately mean, that the distribution of the services shall be on a regional level which would be the broad base planning for a balanced development.

Now coming to details, next to the education comes the turn of the health, which requires to be attended on priority basis. Still in many areas health is looked after by the *tantriks* and *quakes* and from this one could judge the requirement of the bulk of efforts to be put up over there to make them at least understand first the meaning of the health, so that they could adopt the new medical techniques, otherwise what would happen that their reluctance to accept it would be a great humiliation on the part of the Government as its huge expenditure would go as mere waste. They could be given education through mass medias, like TVs, or by other kind of demonstrations, to use medicines for improvement of their health rather then believing in *quakes.* It is desired that at all levels dispensaries may be opened which have facilities such as Maternity Wards, Family Planning Units, Operation Theatres and may have compulsory ambulances and a chemist shop. These are the basic needs. Further near to several rural clusters, big hospitals could be designed, which could be connected to the city by good road system.

The opening of Veterinary Centres are as important as Health Centres because on this depends the inhabitants daily living, as their activities are almost centred around their cattle. A sick cattle to them, is more worriful than their sick child. Now-a-days this is almost a neglected item, with the result of which, many families in rural areas get ruined, as and when, their cattles die due to sickness. Atleast for every cluster, a small Veterinary Dispensary be provided which could be close to the Health Centre and facilities for animals mating and for indoor treatment be also introduced.

The practice of arranging open Bazars is still in vogue over there, where all kinds of items are sold, even now, on the batter system. The Cattle Bazar is also a common feature. But beyond this there is not much in the name of the markets. Those petty shopkeepers, who manage to bring essential items from the cities, usually sell to them at a very high jubilant rates which these poor people have to pay as there is no alternative. Now since, as the needs of these people have also grown, since their economic conditions have considerably improved, so at least a proper market is required for a cluster of three/ four villages which may have specialised items. In facts markets, do not only serve the function of sale and purchase of goods but they are also the common meeting grounds, from the very beginning of the history; as one could be reminded by the Agoras of the early Greek periods. Therefore, as such the placing of the market in relation to villages and cluster is very important in the life of human societies for all social interactions.

In our villages Chopals, were earlier places of public congregation but with the time they have become ineffective, as population is expanding and the social customs and habits are also fast changing. For gathering purposes a Community Hall or Auditorium is required for these areas where activities like film shows, sewing, tailoring etc. could take place which may help in educating masses at the grass-root level. The facilities for kitchen, toilets, stores could be provided there.

For imparting high and meaningful training to the

youths, Polytechnics may be opened in a cluster of 3 or 4 villages so that their employment problem could be solved. A Stadium could be built along it where indigenous sports competitions could be organised. This way the youths could remain active. In each village a Library may be opened and within a few clusters of villages a big size library is required to raise the intellect standard of these people.

These rural settlements are bound to grow and merge with the cities sooner or later and at that stage, there would be absolutely no vacant land left; therefore it is desired that from the very beginning ample open spaces be left inside as well around the village which may act as future reservoirs. Indirectly land may be put to use for temple, graveyards, etc. The planning of open spaces is required to be very definite, otherwise vague open spaces have no meaning; it could be either a mini forest, park, wood land, nursery or like that. From the experience it is being observed that around urbanised villages whatsoever little open land was available, is being now given to housing societies or for other urban uses which is not a healthy sign.

As and when, these areas get urbanised an acute shortage of housing particularly of land less persons and harijans is created for which from the very beginning a reservation could be made preferably outside the village and this could be linked to the old abadi area through perfect blending by open spaces or parks.

The village approach roads are to be properly looked into and they should be made easily accessible from the other neighbouring areas. They may be laid with proper slopes so that water may not stand there.

Regarding sanitation they may be educated to adopt clean environment by providing latrines and baths within their houses with septic tank arrangements and keep them clean rather than depending on community latrines. The community taps may be located at strategic locations near wells and they may be made more hygienic. For the cattle also sufficient measures may be taken to provide adequate sanitary measures like providing drains etc.

One thing should be kept in mind that whatever we are doing good for these areas we are doing it for our next future generation and that is in our interest. For this, all measures be taken, to see the children healthy physically and mentally.

29

Improving Urban Environment through Architects

From the most aggrevated intensively crowded cities where only life-less structures exist like for example New York city; immediately future vision of our less developed cities is foresighted. What an incomparable gap exists between those megapolis and our rural helmets. Those are dreams and beyond that something more is also expected; the extreme fate of the human settlements; distortion and destruction.

From the purest to the most distorted architectural forms have come up the banner of modernism. Those who criticised them earlier speak now in praise of that, despite the fact that the tinch of humanity and socialism is missing in the present practice of architecture.

The fight for space which is haunting large cities has left no scope for the architects to show their talents.

The architecture terminology remains totally diluted; for which the institutions and professional bodies, could be held responsible. The repeated age old norms failed to accept the challenges and as such are stenle in the present day context. The diversifying of the profession from its main axis is widening the gap of its distortion.

The wonders of the world have left exemplary traces of

the rich taste of the builders of the earlier periods; synonimious of which never existed. Every great builder had been crazy for his buildings to look superior to the others of the past. The buildings enjoyed totality and were masterpieces of architecture despite there being constraints in terms of technology in those days.

Most unfortunate part of the building trade today is the vacum created by missing craftsmen of the past; who either left profession or did not care for their hereditary trade for preserving national heritage. Did we ever find any building having columns resembling that of doric ionic, corinth in order or had every building possessed its own personality? Earlier the buildings were the personalities of past builders. In terms of environment, the canals were introduced, gardens were laid, the views, lighting and ventilation were given top priority in the buildings. And so we find the lay outs of the best cities of the world like Mohenjodaro, Harappa, along river Indus which is now in Pakistan. The granary complex was the main building of the city. The houses had open court yards and faced the streets; the latter laid in grid iron pattern. The drains and paths laid were of super order. The potteries and vessels reveal the talents of craftsmen of those periods.

With the spread of industrialisation all over the Europe, the degradation of life style started. The awareness, provided for crisis, was mainly from effluent countries which prospered and subsequently feared from the consequences of urbanisation that became a threat to humanity. They inturn desired that the other developing countries situated in their neighbourhood may also help them in protection of the environment and so were successful in their attempts.

The sea, that joined the world was a cause of concern in terms of pollution and increase in its level on account of increasing carbon dioxide that threatened many countries like

Bangladesh against sinking. The depleting of ozone layer, acid rains in atmosphere, the pollution on account of water, air, noise and wastes continuously haunted the environs of growing cities. the migration of rural folks to towns and cities created the dark slum zones, where existed all kinds of inhumane activities. The fast urbanisation left no scope for trees, open spaces, monuments, canals and water bodies to survive. For the depleted oxygen level, it can be ruled out that the day is not far off when oxygen shall be sold in the open market for survival of human beings. The doctors have reported more psychic cases on account of pollutions and predict epidemics in future.

The urban governments have hardly any infrastructure to help city people. Their budgets are meagre. Hardly any country has so far encouraged voluntary organisations or NGOs. There is no appreciable response of public participation specially in the context of the developing countries.

With the result that urban development scenario is in a total mess. For present no representative of a profession being in the Parliament, the concerned field is totally ignored and if at all some job is done, that is half heartedly executed.

The cities are mushrooming with sky scrapers; the sprawl of unauthorised colonies is found every where, urban villages have acquired form of large commercial centres, the urban fringes are enjoying the city benefits without doing any payment for that and so have developed many loopholes in the city system which have no solution. A time has come when all require proper research and redressal. Every unauthorised colony is bound to become tomorrow's Chandni Chowk of Delhi, India. If so then can the city sustain that? If not then have any efforts been made now? If not then what were the reasons? Have these reasons been made aware to the public? Certainly if public is made aware of properly then voluntary agencies would definitely jump in for saving the future cities from disaster.

The population explosion is being considered as the grass-root of urban crisis. That is true. The concept of developing growth centres in nearby villages had also started but no progress was seen on this front. The concept of Regional Plan, 'NCR' was floated to regulate population of cities, but it is observed that wherever such schemes have come up's more problems are in sight. There is more inflow of public to the mother town. There is more traffic chaos on the city and on peripheries as fast traffic has not replaced the present slow system. So when one aim is thought of its draw backs are not visualised with the result that the aim is lost sight of. The aim, of course is to see also fulfilling the social objectives; the employment, healthy living, reduction of commuters, reduction of infrastructure and less load on traffic. If at all we develop certain system we are increasing budgets for others.

The traffic is the major item of a city growth and more expansive it is, the more we make the city sprawl. Only fast traffic outside (F); more greenery (G) and compact towns (C) i.e. 'FGC' can help in functioning of better towns and cities says Mohan's concept. This will improve the infrastructure and reduce its costs. The infrastructure has to be simultaneously developed rather than depending on our meagre resources of drainage, water and electricity. The gas pipes, telephone wires and other systems need perfection and there should not be any digging of roads frequently.

The sheller front requires more attention. Like Britishers more attention by the authorities could be given towards providing rental housing and hostels for bachelors. That has sorted out much problem and provides social security to the people as well. There is no use of having two faces of the Government; one to control the urban development and the other to work in a manner that control may not be possible. A discipline of a city, country is warranted in which public and authorities both require proper mating. To talk about slum removal and for dislocating their employment avenues is not proper. Are they not the citizens of this government?

The conservation of ridges, rivers, monuments and open spaces is required which would be our lungs when cities alround will have skys crappers and no where will be found any breathing spaces.

Shajahanabad, is India most beautiful example of a city which started in 17th Century by Shahajahan the great, with its lanes, katras, chowks, streets and wonderful that extraordinary effects of serial vision. The traffic chaos and markets congestion has, however, spoiled its functioning which requires proper surgery. The old Faiz Canal that flowed earlier in Chandni Chowk, require revival and so do other lost features of the area. One of the chowks near TownHall requires complete pedestrianisation.

The economic base of a city requires to be duly cared for. The industries, commercial centres must exist but should be like green cities. Why we cannot have it be? Is it written somewhere that industrial areas cannot have lakes, canals, animals, open spaces and trees? Then why not let those be like that as in the 'Donetsk' town of Russia.

The architecture profession is required to take a lead in the field of environment and urban development. The fundamentals at the institution stage are required to be made very clear as to what is required and expected from students. Are they just going to live in a dream world or would like to face the real life.

30

A New Syllabus for Environment

Ever since environmental awareness is created several bans and restrictions are made many of which are sterile as the generations long practice gets disturbed specially in the field of Games and Sports. The dilution or complete removal of bans would be an extensive exercise for the environmentalists which so far has not at all been touched even. Of course now one could go for hunting and revive forests laws. The poachers activities rather than being offensive exercises would also cool down.

It is felt says Mohan's concept that as much difficult it is to go ahead, coming back is equally hectic. This would give a syllabus for environmental protection which is totally missing. A yardstick shall help the people to protect the needful issues and look for the livelihood of many. The later is a must as at many places the entire village or town people are starving due to bans installed bythe environmentalists. In fact this system was very much required from the beginning. If a person has unknowingly killed some birds then it is an offence and he can be put behind bars. Is not that unbalancing attitude that birds killed for Rs. 500/- or say so can give imprisonment for months to any law breaker. That needs redressal.

If the birds of a particular breed already have multiplied

then the requirement of infrastructure is equally important for that which normally is lacking in zoos or zoological parks. Now such kind of problems are being faced but none has dared to go for justice and improve the laws.

Like a student appearing for examinations there is a syllabus beyond which nothing will be asked, similarly the public will have to have knowledge of conserving species and like to concentrate within the said frame-work. Ultimately those would lead to technologies for improvement.

This would provide a good break to people specially in towns and villages whose entertainment activities have all vanished in the wake of environmental consciousness. They would find relieved and feel stronger about their parental activities being revived.

31

Linking Environment to New Technology

A time has come when technologies need fresh thinking that would take up all issues and become an alternative to environment. The various needs and development have to be assessed and with computers, create a solution for the same. Whether this may be on papers or fields, it hardly matters; but a way out must be found.

For urban development issues all problems of traffic, residentia and, commercial shall be defined by computers and even land use of 10 Sq. yards could be found accurately rather than estimation from a micro chart which many times become difficult totally at site. The Master Plan drawings are such papers that need more elaboration and correctness so that correct replies can be given.

Most important is the protecting the people from natural disasters like earthquakes. Nobody will come forward to tell what precautions are to be taken. How should our structures be built. The fast traffic is needed for big cities rather than moving in traffic jams and getting delayed for hours. The protection of rocks and maintaining green forests, all could be a part of those precautions. The preparation of Master Plans too, need such surgery. The existing slum areas cannot to be demolished easily and they require alternate arrangements.

The replanning of areas is needed. In fact the infrastructure for feeding a large population is becoming difficult and a time is coming when every one would have to install his own system for electricity and water supply. The services of Municipal Authorities are very scarce. Of course the cost of living would rise but there is no solution. Identically wastes would be thrown off and incincerated by individuals or centrally by the community.

Even for breathing fresh air gaskets will operate and at places artificial rains would be generated for keepign the dust off. An exercise of do's and don'ts shall be tabulated and for all independent exercises, if done, would reveal very interesting results. Says Mohan's concept, that technologies will completely overtake all environmental improvement issues.

32

Shahjahanabad is Everywhere

The Shahjahanabad will be found everywhere as was in the past. The same would be observed in coming times all over the World as no one can stop that. The basic cause being lack of salaried jobs among youths and general poverty. Every residential street of today is going to be tomorrow's crowded bazar. Some people are fascinated to visit such places off and on where as some don't. The life style is the same whether they are at Bangladesh, Pakistan, Egypt or in India. However, for western countries with sparce population that may take time but in Asian context concept of streets and narrow lanes cannot be averted.

All unauthorised colonies, the urban villages and the slum settlements will in particular be a part of this system. In case no efforts are made to provide facilities there, right at this stage, then no one can save those areas from the disasters of fire, stampede and over crowding. The people would no doubt have such areas in great number but at whose cost? The human values!

The author feels that none has come forward to create any awareness in this respect as everyone is bothered about the present state of affairs and as for as the future generation is conerned, where would children go for schooling, playing and for employment, that seems to be no one's business to think.Is it that our next generation in toto is going to be absorbed in the commercial sectors and in the name of green lungs, perhaps then there would be none.

The merits of living along streets pattern and shopping cannot be overlooked but at the same time an awareness among masses needs to be created so that future constraints need incorporated right at the initial stage so that these may be comfortable areas of living. The demolition of such streets we know cannot be done and where is the man power and official machinery to do that? Till the three faces of society i.e. the public, bureaucracy and politicians survive those patterns of settlements would be there and survive much better than the planned areas.

The intellectuals may like to live close to such settlements or areas where they enjoy light snacks during evenings, have a rickshaw ride, see people and come back home satisfied and full of new ideas. Is it not surprising as to how a large number of people living in cities do miss such economical facilities of entertainment.

Every object in the World after a generation gap acquires the sense of love, attachment, monumantality for which with the human settlements aspire to either be with those, may be atleast for a few moments. From the refugees in India who came from Pakistan after the partition their emotional talks always remained oriented towards their past residential premises lost in some street of Lahore or Karachi where they spent their childhood, playing kites and running under the sun in the narrow lanes. Vis-a-vis the refugees in Pakistan, who left all their belongings in some streets of India must not be leaving any Indian visitor go without enquiring about the area where earlier they lived. Those streets still, are living though, the inmates have changed.

The governments on both sides have in the past made efforts to allow the tourists population to allow visitors to visit the two countries yet that much freedom did not exist, so none could enjoy a healthy scenario due to the restrictions of politics and security both.

The streets of Shahajahanabad lost many such residents who could never return. Their relatives left behind are believed to have written earlier few letters to them but those sentiments too have become diluted with the time and the memories ultimately got vanished. Many expired also with the time.

However, those streets have become popular not in the country only but throughout the world. The younger generation occupied key seats in the business of the elders and young brains, though, brought tremendous economic revolutions but at the cost of social upliftment. That kind of society which existed earlier among which the entire city people were family members was no more. One remained confined to individual family only.

The areas of sparce population with single storeyed structures became intensively developed in the multistoreyed complexes leaving no space meaningful on the ground as a result of which the streets became narrow and crowded for any vehicular traffic to move. The hetrogenous traffic both slow and fast plied on the streets of the area.

Every street enjoyed an identity of its own with a speciality, worth mentioning of which are Chandni Chowk, Maliwara, Ballimaran, G.B. Road, Motia Mahal, Lal Kuan and Sita Ram Bazar. All remain interlinked with a net work of other smaller lanes and by lanes all framed up within the walls, with all facilities of mosques, madarsas, hotels, restaurants and gardens created in that.

One who has not seen streets of Shahjahan's cannot be expected to have visited at all Delhi, the capital of so many earlier rulers—the Hindus, The Muslims and the Britishers.

The Britishers seem to be least fascinated with streets concept and often rejected this old settlement the glimpses of which are found in, while Planning New Delhi that ignored totally Shahjahanabad. However, the tourists always liked to visit those areas. The streets have a combination of many innumerable institutions where the heriditory jobs still persist though, the business may not be so profitable now. There are examples of dying arts where the uneconomic situations have prevailed and which is forcing the professionals to change their titles.

The great poets Ghalib and Zauk, who recited their poetries in those streets and one could imagine the audience listening to them. Still such streets of the walled city are rememberances of the past, and famous amongst them is the Ballimaran where lived the poet Ghalib, the greatest urdu poet of his time. The important personalities under whose name the street names or chowk names were put provided a distinct identity to the area.

A distinct communal feeling existed in the area that draws at times the scene of brick bats throwing and such untoward scenes which are part of the area that can never go. Those are very sensitive issues and the author also feels that whenever he visited such area s he has remained very careful.

The easy employment has attracted many institutions to prosper and from jhalliwalas to rickshaw pullers fall under such categories supplemented by other institutions of cheaper *dhabha* owners. The M.C.D. Slum Authorities provide economical accommodation to the people in Night Shelters built in the area.

Many visionary links are attached with streets. Sometimes it is dark to light zone and hot to cold. The streets sometimes open up into large areas called Chowks which again provide a direction to some other street. Besides chowks there are excellent land-marks in the area like phataks, gates, Kuchas,

Katras, Havelis and Chattas, synonimous of which are not found in the world over. The area has been covered up technically under a special area plan and various policies of Mixed Land Uses, Conservative surgery has been advocated for different parts of the city.

The conservative surgery aims to provide mandatory green and parking lot in the area with the same F.A.R. The Mixed Land use allows few streets to remain commercial. There are specific proposals for each area wards. The M.C.D. is looking for implementation of approval of plan as per land uses infrastructure like sewerage, for the provision of dust bins and Dhalaos. A proposal recently came in press that a U.K. based N.G.O. solved sewer problem of such areas. So those espects could be referred to improve the area. The conservation of monuments in the streets are a special feature for which archaeological deptt. is also interested and no building within the 1 km. area need be disturbed. There are beautiful monuments of Jama Masjid, Red Fort and Fatehpuri Mosque which are available in the history records of almost every old historical book that need be looked into. The Brtitishers tried to ignore that and spoiled our heritage. Even other rulers who invaded the city among which Nadirshah is one, much harm had been done to the Monuments. Even then each movement required extra care.

The Police and Fire Brigade Role in such crowded areas becomes excessively difficult. There has been lot of demand of setting up of Fire Stations and Police Posts but due to lack of space that is not possible and nothing better can be evolved. Things could remain as they are.

Not much could be done except to remove the congestion and get the traffic conditions improved. The law and order and saving from disasters like earthquakes, fire or such ıntoward incidents have never been thought of which could be possible only when the wider lanes and streets kept open all the time to avoid any stampede. No one ever made any research whether business houses could also be restricted to help employees save energy and avoid rush. More green area is required for which one has to pay the cost. The Slum Deptt. of MCD may help in the mission if the dilapidated buildings are pulled down and green areas provided.

A group of the wholesale professional trade could be provided massive alternative accommodation in some other parts of the city as decentralisation policy, would help in shifting out its population.

The pollutant industries have already been removed as per latest Supreme Court orders. For the pollution of air, noise and water people are quite conscious and more alert. The traffic situations every one desires to improve but how? The slow vehicles could all be replaced or certain alternative arrangements be made.

A strategy for vehicular segregation could invite a few creation of plazas near important chowks at +.75 m level and proposing parking at -1.35 m. The second strategy would be to have a 3m wide corridor 1.5 meter on either side of central verge for slow vehicular traffic at +2.85 m level. This experiment could be used at wider locations and if successful may be attempted at other chowks and streets inside the walled city.

In fact had the Britishers taken proper account of this city and had there been a great Shahjahanabad plan prepared in place of planning New Delhi then perhaps things could not have been so much spoiled. The great Shahjahanabad plan even was later ignored while preparing 2nd Master Plan of Delhi in 1962 which surprisingly praised N. Delhi plan of Britishers. However, the plan could have extended a few km. upto the river and in other directions to accommodate even a university complex alongwith a large size hospital besides other required community facilities.

A fresh approach in terms oof providing cosmetic treatment to the city at the time is required so that pollution aspects be minimised. Everyday one institution be examined for decentralisation. The slow vehicular rickshaw traffic need altogether be minimised to 10% of the existing number. The trees may be replanted at places required. Let the tourists come there to enjoy at each and every spot and let them fix a route to be destined with suitable guide facilities appointed. Let the entire streets of the area be monumentalised and preserved.

The other few areas like Qutab Road, Paharganj, Sadar and Baratooti which are major markets and tourist centres though, form, extensions of development, need improvement. One of the British planner accepted Paharganj to form a part of N. Delhi when the concept was given up. Now it attracts major tourist population of city in its lanes and bylanes.

You will still save
the building of future
Chandni Chowk the congested street
with jammed traffic
and the curfews
as well as fire accidents

By your little efforts
still, go to open
where urbanisation is in process
guide them to have
better environs
just with little efforts

A model of good living
be telecasted
and be made popular
that is what you can do
for humanity and
people of tomorrow.

33

Mohan's Concept

Mohan's concepts have been framed at Mohan's Urban and Environmental Research Centre (MURC) and approved by the author.

1. The Master Plan needed technology improvement and shifting of all residential/commercial uses from the centre and move them to NCR areas.

2. In privatisation, strictness is desired to be observed by Govt. Deptts. First of all, it is desired not to give the full job to one person. Secondly, the jobs to be allocated and time limit be fixed with amount of fee to be paid. It is realised that one time payment for work and Quantum of job within a short time can bring results, otherwise Govts. officers can create problems.

3. The one Indian Cricket XI team now requires re-orientation. It is necessary to have even 10 teams and let in the World those any may move. Then there will be no serious thought of fighting. Identically all professions require multiplications solutions so that work could become easy.

4. The 'Rural use zone' suggests a wide diversity in terms of rural uses. But one has to omit the rural impression from the minds of those in Delhi. Those are many

backward areas and from the concept it emerges that they are the targets of law and judiciary and this combined with poverty makes them suffer. Earning two meals a day becomes difficult and they do not want more so as to be distributed. They may not like to come in focus as to suffer from local riches. If you can introduce good finances for them and make them reproductive, that could be better and save immigrations.

5. Often this is the talk of the senior official's in which one wants to exert that he had been good at. But Itell you that those who took interest in physical planning of Delhi, destroyed that. Even up till date nothing like environment feeling could be effective. The thinking infact was guilty. A total scenario could never be conceived of. The expansion of the city size and urbanisation is one factor. Did we really need that? That helped or destroyed Delhi? In fact the technology adoption was never used as a tool for improving the city which is required for a growing city. Avoid intensive urbanisation and make compact planning as and when required. This is the missing first link in the complete city growing process.

6. The green belt restricts the city growth but you don't know from where public would enter and build. The joint schemes get bus facilities and other more easier facilities than others to develop. I think one has to restrict to facilities and create them in case we want to protect cities. Moreover these had produced economic issues so a continuous monitoring is required to be done.

7. For a project the development should be compact with all facilities. There is no need of fast urbanisation. What is the hurry? How do we provoke the clean

environment? An exercise be framed up for an area specially during hot summer when driving or walking looks awful. In the market eight or nine such plans need be displayed for adoption of any one for the developed work. Time to be taken, the plans to be adopted for years, need study.

8. The students feedback system is superb and that too without any name and address mentioned addressed to atleast five heads so that the concerned teacher's weakness could come to light.
9. Like 26 January in India every village need celebrations. To hoist coloured flags all round would look elegant.
10. For helping the youngsters appearing in final exams, I as an examiner never could say 'no' but had seen and passed by lest the students get low marks that may stop their entry into good courses. Rather I gave them such lavish marks that all got good divisions and took up good courses and were placed well in the society.
11. In a good size park an exclusive area be earmarked to house cages of animals and birds for daily or week end visits for mixing with animal life. This way, every day, cages, could be carried to different parts of city from Zoos. This will create more employment and money flow. In a budding town the parks require building in advance out of landlords money otherwise later on there is a problem.
12. Every one has talked of the protection of the ridge or the river. No one so far has said anything that addition to the same is also valid. Mohan's concept feels that rather than scratching the natural green for putting to urban use, the urban use need be dissolved in that. For example for Yamuna river bed even Ring Road could be sacrificed and master plan changed.

13. The Vendors selling Khichri, dalia and other health conscious items should be given incentives since pollution has created a problem of health for the city people. Even they can possess exercisors to keep people healthy. They may even go on pheris, moving stalls selling them.

14. Much has been talked about the environment. Now let's examine the repercussions of the same and describe a specific syllabus and what we require for the same. Or else now we may even do away with that and contrary to that come about anti environmental issues.

15. Do we really require anti environmental talks? Yes the time says so. A syllabus is definitely required as also what items are kept for the scope of environment. The lefts may have an another slogan for environment.

16. There should be more then one professional institute like that of AIIA, AITP to have competition and better working. Like NGO's all should act and have active participation in the welfare.

17. A development project belongs to major tourists where every corner for tourist interest need be created to bring better remuneration and an improved culture of exchange of goods and business. All ponds, wells and temples in the area need be linked to rails and roads. A good site, bank, market facilities be created which automatically would become intensive. The norms for ponds, wells be made and sanctions be granted for minimum expenditure on them.

18. The Architecture Planning professionals should be all neutralised by converting in to living or life. The bachelor of life and master's degree need imparting all issues relating from birth of a person till death.

19. Mohan's Concept followers if they want can become Mohan's fellows and a recognition of that could be converted in to a University awarding certificates recognisable as degrees, post graduation degrees and doctorates of life.

20. Every one says you are good but often refer back that so and so is bad. Life is like that. What Chankaya said to his mother that father remains in an other word and does not face reality. So the later part for consideration is important to realise this world.

21. What do you think is employment? This is healthy living and smooth functioning of life with wife and kids for shelter and bread.

22. Any profession which does not reflect the service to humanity and to self is useless and baseless. All profession at the extreme are same, truth, and nothing beyond.

23. The professional course requires to be banned. Those are spoiling our social relationship's on the grounds that those passed out become egostic. Let the general BA, BSc courses have a special one year course maximum as professional.

24. As driving is becoming necessary and with traffic jams, such time is spent and more tension prevails. Some drivers driving good cars with ego are driving in tense situations. Those require relieving for those hours by switching on to radios to listen a particular programme.

25. The planning needs focus on orchards, ponds, temples and railway lines. There is a need to orient green paths, orchards, gardens running parallel to them. There is a need to combine music, dance schools and institutes to match with green spaces.

26. The Author is the 1st person who in his 1st book "Environmental Awareness and Urban Develoment", in 1988-89, introduced a comparison of India's works with global strategy. It was after that the news papers mostly from India introduced a separate global news page.

27. The author in this book has tried to adopt flashings of articles as is in cinema.

28. The private enterpreneurs having Commercial, Residential plots could develop ponds, water bodies as facilities. The parking could be the problem. The technology is facing difficulty of parking. Why not then cars to be converted to small size shopner type kept in purses, to be taken along with every one. Is that possible? Yes Sir, how? to be looked up by researchers.

29. The urban attitude especially by development authorities is very tough, like a devil that requires modifications. The understanding between a developer and user require depth and proper meeting. Do it or don't do it, policy requires modification.

30. For the future of unauthorised colonies, the Mohan's concept reorients the studies of urban planning towards purely being investive in terms of futurisation. An unauthorised colony after years, say 20, becomes a slum. So what is required now is an intensive slum area that could be introduced there which could earn good business requires rethinking.

31. There is a need of parks, open spaces, community centres, wide roads and good shops dealing in most descent items. The best is to develop private parks, which could charge the users. That could be for children and for aged a club where one could sit with

descent gentry or places particularly for indoor sports as well as for out door. Even the Banquet Halls are needed. So the plots are taken for that use that could be helpful for the public. The banquet halls planning could be more useful.

32. The private parks scheme require more popularity says Mohan's concept. Along canals, rivers incentives be given to private orchard owner's to develop private parks.
33. The age old hawkers are of special significance to whom the author would like to meet and eat the delicious items at much cheaper rates. The author personally recollects his school hawkers selling items of his taste, batise wala, kulchawala, kachaluwala etc.
34. Be aware of retarders. Those can never be removed. Those are serious pollutions.
35. I think there will be no Master Plan effective for Delhi. It will be a plan of improving technology, improving present systems and keeping every other issue as those are.
36. The flash system of introducing cut and paste technology of medias could be introduced in Planning also while writing article of reports. The fancy cities look better. How that is to be done is to be seen.
37. The CRASH courses in architecture and other fields may be opted. However, option could be analysed as:-

    Do we really require lengthy course of 5 years?

    No/Yes

    If no, then how much?

    Three months?

    How come?

38. Never can we replace traffic in markets with pedestrians, otherwise that loses the charm. In a community we should have a competition; tournament spirit that will keep all healthy. Like in blocks games for cricket, hockey will be appreciated. For ladies cooking can be competitive.

39. The city's structure should be one single small size square piece say 4000 x 4000 yds and relationship be thought of by creation of forests or by important links. Conceiving a city of large size is not possible, nor feasible and is not required. The design of a square must relate to the khasra nos. on the ground so that land issues could no more remain a problem.

40. The author was the first one who introduced the concept of total world related to individual, in his first book "Environmental Awareness and Urban Development". This was picked up by newspapers later on. Because there is no work which need no comparison and competition with that.

41. The entire project is a sort of exhibition of a house belonging to one person and the later may be in a position to show the house to all including details.

42. Since the ban on animals and birds for keeping them anywhere an artificial surgery is required by Artists, sculptors. From the old movies we find that the head of village jamindar or rulers has a Tiger full size (dead) kept decorated in his bungalow where English people drank by the side of that. The artists in a way can produce dito copies of that to impress upon. Even paintings could also serve the purpose. A colony or project area can also have that. This can be helped by professionals who prepare effigies of Ravana in bulk. At Narela also in Delhi we would like to have images of Tigers, peacocks, parrots, and let those help in

forming landmarks of the area. The Tigers marks reflect mothers attitude to a city. It is better. Then at place could be placed 5 Tigers, 2 Tigers, one Tiger to provide different landmarks.

43. The applicability of Mohan's concept started from specific professions of architecture and for planning and gradually spreaded in all directions of truth since every profession is related to truth for man kind. Therefore, the justification of this is concerned with family life, individual good, good employment feels Mohan's concept, that if those suffer then planning and architecture cannot be spared for being accountable.

44. The project planning has become the general issue these days as no management feels to sustain loss financially at the end. Therefore every date pertaining to complete a job is fixed. When so much sensitivity of fiscal issues bothered, then, ecological issues have hardly any weight as the least fixed measurements are allowed to get the project a green signal.

45. What ecologically do we require? As a matter of fact the wider road beyond 30 m wide 1st of all needs a touch. Can that road have no pollution of vehicles and can the public move on that? Have we provided long corridors of roads with pavements. In Bhopal, India we could find long stretch of roads on which the public is found strolling during morning and evenings. The other points in brief are mentioned as:-

    1. Avoid taking up site for creating environmental problems.

    2. Trees maximum should be on west side and that too of fruit varieties.

    3. The frontages should be welcomed with greens.

4. All water bodies/monuments need be 1st tackled.
5. Linkage of religious area and ponds need be created.
6. No slums or unauthorised construction be permitted in the area.

46. Environment consciousness for a New City should have a major core and space to expand further. Mrs. Salma Aghahas rightly sung in "NIKKAH" film "Hum ko jobhi dost mile bewefa mile." That is what city issues also refer to.

47. After environment What? and what would be the after issue of that. Is that goingto last very soon the issues on environment? First of all let's find out how this started, who favoured that andwhat was the need of that.

48. For the married people training programmes need be made for better family management keeping for good health and for maintenance of the house.

49. For the old subsidised canteens be run as well as medical aid need be provided and religion oriented trainings need be worked out. For health trainings few more points are as:-

1. It is better to remain anonymous than to be known. Freely roaming and non-bindable person is more free.
2. The child psychology or women's attitudes could be best worked out by watching them daily and search new changes in them.
3. It could be more diversified by watching at each time/every minute and second. This invites a number of presumabilities to be worked out in detail.

4. The dens of love should sparkout in every building and environment closed or far off for one to enjoy. Two colleges ran earlier in walled city, Hindu College at Kashmere Gate and Delhi College at Ajmere Gate. There must be certain relationship formed. The Delhi College must be very popular as was close to muslim population.

50. Delhi has been crushed from time to time since 17th century and especially after 47. Each administration did its job in a bureacratic way. There have been layers of destruction laid below its growth layers among which Raj period is more responsible.

51. The fire officers may be given the staff and sanctions for providing relief to the masses where such accidents occur. The establishment cost need be sanctioned to them. From time to time they need circulation of the papers.

52. The concept should provide facilities to all categories of people, children, youths, married people and old. A few other points are :-

 1. The children development activities need proper orientation.

 2. The youths be informed to get rid of the careless attitude or jobs oriented programmes and sufficient systems be developed for them. The sons, daughters marriages bureau problems be sorted out.

 3. There is no need of permanent allotment of land to any one. The vehicular facilities could be provided temporarily to carry out daily activities.

53. It is necessary for an individual to learn each and every issue. For example initially for architects, planners, 12 training courses per year may be made necessary of different disciplines, stores, health,

economics, social, arts and crafts, horticulture, security etc.

54. Every meeting before starting the speaker's representative must in brief exercise the audience about their knowledge on the subject. Then that text he may start.

55. Even huts which are in jhuggi cluster should have space for latrines and baths which may be very small. The community latrines are the worst and above that too paying fee for using the same? Even middle class families may have difficulties in paying for using toilets.

56. Every zone comprising of a population of 10 thousand people should have a force of approx. 1 per hundred persons to look for any fire, floods, earthquakes and to report within 5 minutes after physically going to the spot. Such NGO's may be paid high remuneration from Government for serving the public.

57. It is very strange that a body to whom (NGO) a project would ultimately go, hardly peeps earlier into that saying that its jurisdiction or accountability has not started. How later then it can handle the wrong things done?

58. The best part of a city informer are those elements that reach the roots. For environment to reach deep interiors income tax deptt. can definitely help. Even concessions could be declared for those who adopt such measures.

59. It is essential beforoe ending how the role of a professionalist is required to meet the environment needs satisfactorily, in which sphere and by how much? I could not see this everywhere. There are thousand of professions from a house lady doing house job, to a person sailing ship on the sea.

60. The role of businessman is also important in the sense that it forms a large part of the sphere. Right from what should we eat, stay or spend during the day is important.

61. Even safe measures are required to be sorted out. This is needed for population control and happiness of family and positive jobs.

62. The floods are a menace to the public after the rainy season. To save the public from this we require a new technology that could help people from such calamities. The technology should be such:

    The water may be converted to Ice or else be made to vanish by laser beams. The Ice freezing within an hour can be made possible and can be kept in over-hangings in large stores over the river till safety time.

63. For cloud bursts that occur generally on hills where all cattle, people, houses get drowned, the drainage arrangement be made for as from the historical evidence. Such areas be checked. Again high routed laser beams could be used to vanish all the water.

64. The Forest fire does also require high class technology to control the same. First of all it needs proper identification, secondly the direction of fire be made known and thirdly the escape routes be known earlier through equipments or else the underground tunnels be taken up in all the four directions for escape routes having oxygen mask facilities and fast escape.

65. So in all such issue we require as:

Problem	Technology to solve	Demerits of that	Cost factor
The storms on sea	Life can be planned	do	do

66. The Author's son Abhas Mohan, 11 years old has an other scheme for solving floods, storm, cloud burst that is to pass a current through river to convert water into Hydrogen and Oxygen. Thus he says alongwith other means we can change floods by digging rainy wells.

67. The late Professor K.D. Mistry of School of Planning and Architecture stated to make the drawings very clear which is missing from architecture and planning side. The land uses are told approximately as the scale of drawings are so small. Somewhat Wilson Size Sheets Scale are advocated for all master plans proposed. Professor Mistry was even told to write an architecture drawings, meaning transparency and clarity in all disciplines. Even tiles laid,its size,colour quantity and price should come up before laying.

68. The courses for architecture should be of maximum three years duration and be changed to Human Settlements with specialisation for 6 months.

69. The slum should be in one voice by public, bureacrats and politicians.

70. In a biabary jungle if roads are tobe carried than the proposal shall be on stilts so that all would be temporary without harming the ground.

71. To a school boy of 6th standard the author's posed a problem of technology that how in a second you could cross a 2 km bridge over Yamuna at Delhi near Laxmi Nagar. A colleague Mr. Shashi Kant justified that and came to a conclusion that could spare the parking problem.

Conclusion

The environment topic has covered up a vast canvass and the laws framed for its protection are at some times not required. From the conservation and urban environmental issues there is a very little being thought of its procurement. Let me know that why those earthquakes are so much damaging the public life when already alertness is given. May be we may think of different sciences to face disaster contents and think in terms of requirement of technology changes only.

The environmental links need clarification. In whichever fields not much could be taken up individually. The marine science is to be more looked into as three-fourth of Globe is occupied by sea. How that is to be vigilant may have to be checked by executing work exercises and public awareness.

The forest fires, cloud burst and tornado can be more disastrous even if work exercise are taken up. Japan has done so much in earthquake technology even then those countries are not so fully earthquake safe.

Much has been talked of conservation of trees, monuments, animals, biotic life and all those need a new direction as the author thinks that gradually, hunting and other aristocratic games could be allowed again on time-table basis rather than being stubborn and not allowing.

The hunting a such past games revival and removal of ban can yield better crop of the people, rough, tough and healthy. The forest life can be revitalised. The poachers activities will be minimised.

For conservation of monuments more cleaning is required and good tourist approach is desired. It should be revised and more areas should be covered up from the villages and towns. That is the heritage of a country that need exploitation. More studies will be required in detail and those would be carried out from micro to macro level. For much time the trades and employment have suffered so putting removal of bans on issues will be revitalised and new life will pour in, says Mohan's concept.

Health of Architect to be kept

The health point of the workers, architects and other will be of vital importance attached to the same institute to guide. The trackings required to be in this direction will be required. At all stages it will not be necessary to go to big hospitals but the Institute itself shall look for that. The precautions for lights for eyes for keeping health shall be required.

Instructions

1. Better to have it edited once more since it has taken much time to write.
2. All abbreviations mentioned better to be converted to full form like

   S.......and

3. Bibliography and index to be found.
4. Mohan's concept papers have an additional sheet.
5. The project mentioned has a single drawing to be read now.

Mohan's Concept

Quick Analysis of any Problem

How much should be the urbanisation? Before dealing with the subject let us study its merits and demerits.

Merits:

1. It looks nice to stay with urban facilities.
2. Availability of good specialists, Doctors doing studies.

Demerits:

1. The infrastructure is very poor.
2. The poor roads, street lights, collection of street water,

poor drainage, non-availability of mass traffic and transportation are the physical problems.

3. The lack of family environment, sociological depression, unanimity in environment, adulteration in food are other problems.
4. More office, houses, poor finances with authorities to develop main problems.

Assessment:

1. Of course it is better to have urbanisation.
2. The Authorities should develop such areas only as could afford only with even provisions of providing AC Cars, houses and all facilities and in case it can't.

Bibliography

Babylone is Everywhere by Wolf Schneider - 1963. Holder and Stoughton Ltd., Oxford.

Cities of India by G.W. Forest CIE, Metropolitan Book Co. Pvt. Ltd., 1977.

Delhi, Its Story and Buildings by H. Sharp,Humphery Melford, Oxford University Press.

Delhi Gazetteer, Delhi Administration.

Delhi - Development Change, 2000, Ashish Publishers.

Delhi, The Deepening Urban Crises edited by Patwant Singh and Ram Dhamija, Sterling Publications, 1989.

Delhi Through the Ages, edited by R.E. Fry Ken Berg, Oxford University Press, 1986.

Design of Cities, by Edmon N. Bacon, Thames and Hudson Ltd., 1967.

Environmental Awareness and Urban Development by I. Mohan, Ashish Publishing House, 1988.

Environmental and Habitat by I. Mohan, Ashish Publishing House, 1989.

Environmental Crisis— A Socio Eco. Perspective, Mittal Publisher. (under press) by I. Mohan.

Ethnicity and Urbanisation by K.S. Nair, Ajanta Publications, 1978.

Environmental Issues & Urban Development of Walled Cities by I. Mohan, Mittal Publishers, 1992.

Environmental Issues: A Development Concept (Slums) Poverty Allertation by I. Mohan, Anmol Publications Pvt. Ltd.,

Environmental Pollution & Management by I. Mohan, Ashish Publishing House.

Forts of India by Amrit Verma, Publication Division, Ministry of Information and Broadcasting, 1986.

Fragile Environment by I. Mohan, Ashish Publishers.

Great Buildings and Monuments of the World, Minerva Publications.

History of the World, via count, the Hamlyn Publishers Group Ltd., 1985.

Homes, Towns and Traffic by John Teltlow and Anthony Goss, Faber and Faber, London.

Inside the City by J.A. Erson and B.P. Fitzgerald, Longman Group Ltd., 1972.

ITPI Worskship 15/12/1989, *Garbage Collection and Disposal System Public Private Co-operation* by Girish Mishra.

Linking the Two Ridges in Delhi by I Mohan, *Hindustan Times,* September, 1985.

Man's Struggle for Shelter in an Urbanising World by Charles Abrahim the MIT Press, 1964.

Master Plan, Delhi, 1962, 2001.

Muslim Cities in the Later Middle Ages by Ira M. Lapidus, Cambridge University Press, 1967.

Planning for Conservation edited by Roger Kani, Mansell, London, 1981.

Planning the Hill Habitat by A.K. Dev and S.A. Rizvi (Eds.) Mittal Publications, 1991.

Planning New Cities on Walled City Pattern, by I. Mohan Civic Affairs, Dec., 19989.

Planning Shahjahanabad by Om Mehta, *National Herald,* 9.3.1975.

Population, Growth and Land Use by Cohn Clark the Macmillan Press Ltd., 1967.

Rebuilding, Shahjahanabad by Jagmohan, Vikas.

Reshaping a City by T. Breman, The House of Grant Ltd.

Shahjahan's Delhi : Past and Present by H.C. Fanshire, Sumit Publications, 1979.

Sick Cities by Mitchell Gorden, The Macmillan Company, New York, 1963.

Slum Clearance and Improvement (Papers) Seminar IIPA, Delhi, 1979.

Slums of Old Delhi by Bharat Sewak Samaj.

Social Life in an Indian Slum by Paul D Wiebe Vikas, 1975.

The Concept ofTraditional Bazars, by I.Mohan, *Design Inc-Indian Building,* Aug. 1977.

The Dismal Record of the Archaeological Survey of India by I. Mohan, *Design Inc. Indian Builder,* Dec. 1979.

The Exploding Cities, Peter Wisher and Rose Marry Righter, Andrel Dentish.

The Forts of Rajasthan by R.L. Mishra, Kutir Parkashan, 1985.

The Future of London by Edward Carter, The Penguin Books.

The Nature of Cities by L. Hilber Seimer, Paul Theobald and Co., Chicago, 1955.

The New Capitals, of India, Pakistan and Bangla Desh by Sten Nilson, Printed in Sweeden, Student Literature, London, 1975.

The Meideval Town by Fritz Rorig, B.T. Batsford, London, 1967.

The Urban Future by John N. Jackson, George Allen and Unwin.

The Urban Prospect by Lewis Mumford, Sector and Warburg, London, 1968.

The World of Walled Cities by I. Mohan, Mittal Publishers.

Town Planning in Early South India by C.P. Venkatarama Ayyar, Mittal Publications, 1987.

Town and Town Space by Thomas Sharp, John Murrey, 1968.

Town Space by Gorden Cullen, The Architectural Press, London, 1961.

Tropical Architecture by C.P. Kukereja, Tata McGraw Hill Co. Ltd.

Urban Development in the Third World by John D Herbert, Praeger Publications, 1979.

Urbanisation in Assam, by Deepali Barua, Mittal Publications, New Delhi, 1992.

Urbanisation in India : A Case Study by Marirul Huda, Mittal Publications, New Delhi, 1990.

Urbanisation : A New Outlook by Harshad R. Trivedi, Atma Ram and Sons.

Urban Squatter Housing in Third World by Ashok Ranjan Basu, Mittal Publications, New Delhi, 1988.

Urban Studies, edited by Louis K. Loewenstein. The Free Press, New York, 1971.

Urban Studies, A Select Bibliography by P.P. Singh and Ramesh Pandit, Mittal Publications, New Delhi, 1988.

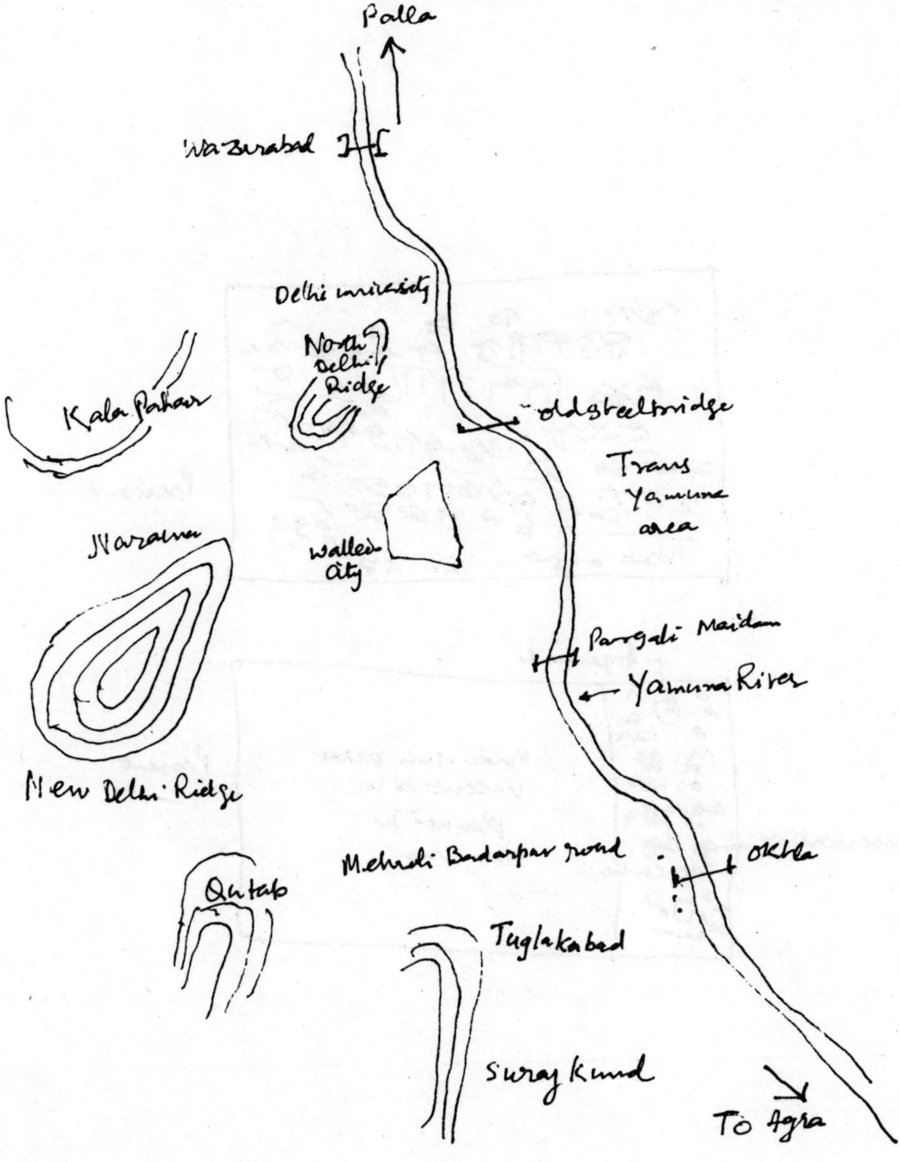

1. Plan Showing Aravali Ridge and Yamuna River at Delhi.

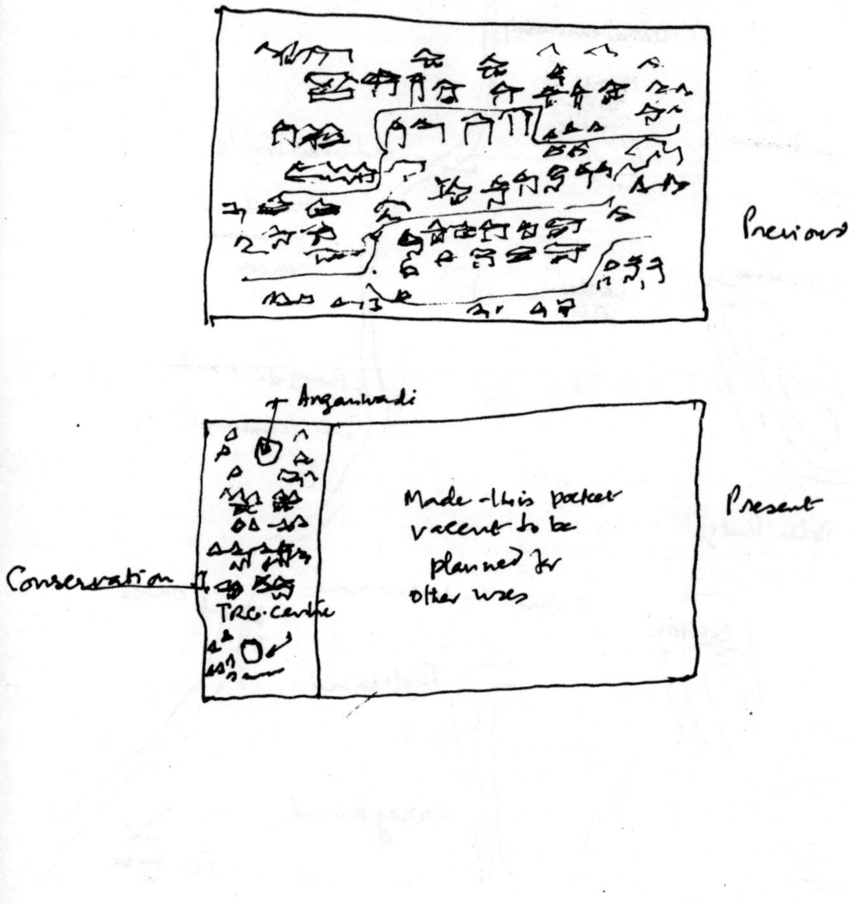

2. Conservation is the Best Mode of Planning an Slum Area for Poverty Alleviation Programme along with an Anganwadi and Training Centre.

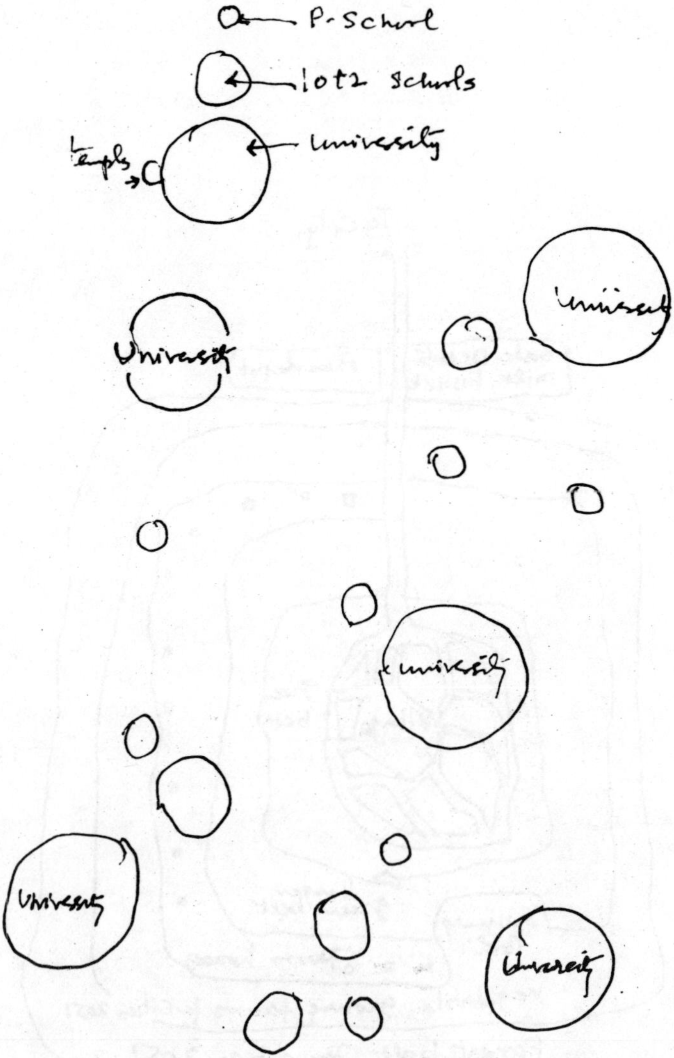

3. **Developing and Linking Primary School, Secondary School and University as Tourist Centres.**

Each one of them to have temples devotes to all four religions. All around enough infrastructure needs provided for particular two days; Saturday and Sunday. Also thought to linking all universities is to be worked out.

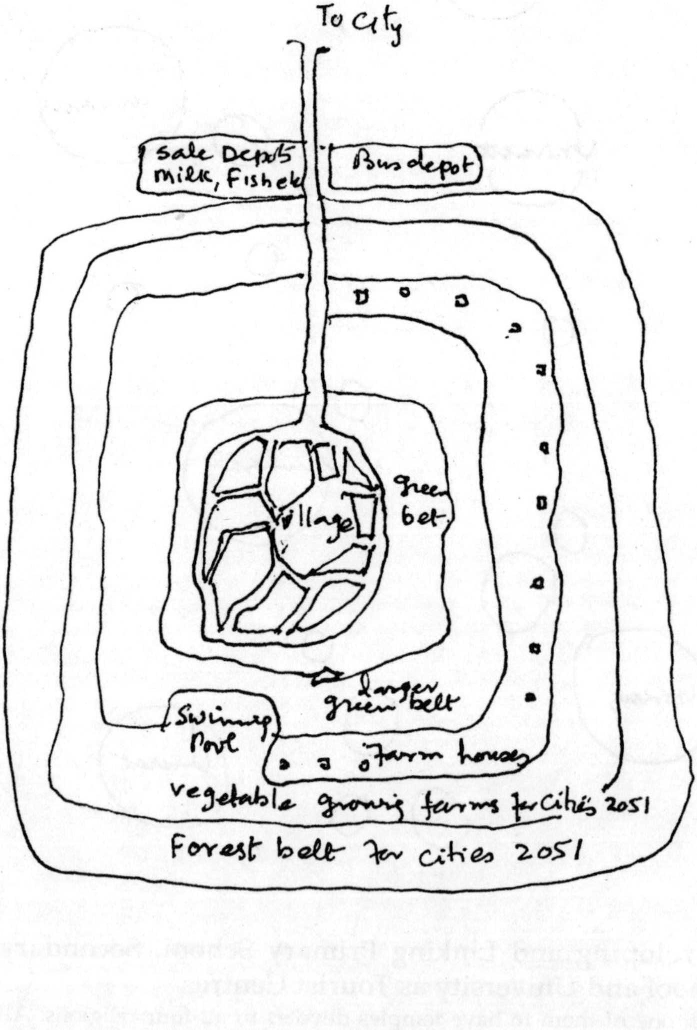

4. Linking Future Cities with Villages.

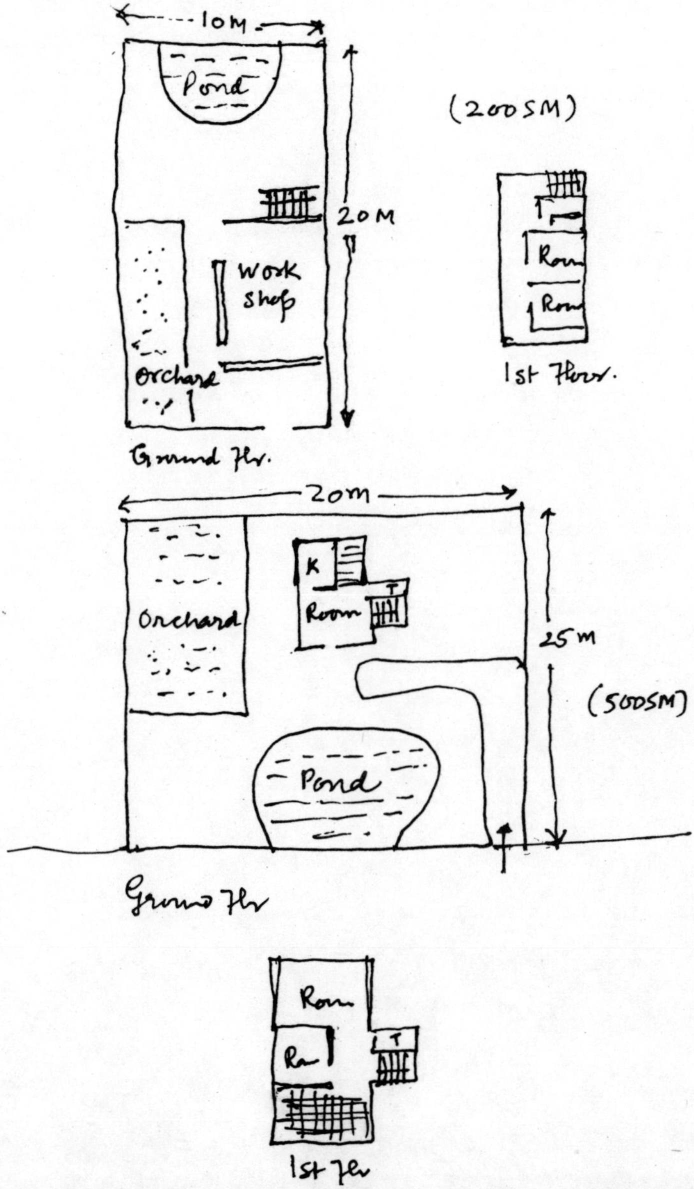

5. Typical Village House Plans for 200, 500 sq.m. Area.